This book is to be returned on or before
the last date stamped below.

PLEASE TRY TO AVOID
RENEWING BOOKS ON SATURDAY

D1346064

COAL

COAL

A Pictorial History of the British Coal Industry

D. Anderson

DAVID & CHARLES
Newton Abbot London North Pomfret (Vt)

To my eldest son, John C. Anderson BSc, CEng, JP, Mine Manager and fifth generation mining official

British Library Cataloguing in Publication Data

Anderson, D.
  Coal: a pictorial history of the British coal
  industry.
  1. Coal mines and mining – Great Britain – History –
  Pictorial works
  I. Title
  338.2'7'209410022   HD9551.5

  ISBN 0-7153-8242-X

Typeset by Typesetters (Birmingham) Limited
and printed in Great Britain
by Biddles Limited, Guildford
for David & Charles (Publishers) Limited
Brunel House Newton Abbot Devon

Published in the United States of America
by David & Charles Inc
North Pomfret Vermont 05053 USA

Contents

Acknowledgements

I am very grateful to Mr Joe Gormley OBE, for reading the manuscript and to the following for the gift of and permission to publish photographs: Mrs Phyllis Lane of Harwood, Bolton, widow of my old friend Jack Lane OBE, Divisional Inspector of Mines & Quarries for the North West; Miss Sally Walker of Highfield, Wigan; Miss Mabel Ashton of Garswood, Lancashire; Mr John Barlow, Chief Archivist, NCB; Mr Alan Davies of Atherton and late of Salford Mining Museum; The National Railway Museum, York; The Historical Monuments Commission; The Rev W. Bynon, Vicar of Highfield, Wigan; Messrs Sotheby's, London; Mr James Peden; and my cousin, the late James Beckett, who was an authority on mining history in the Oldham district of Lancashire and wrote a series of articles on the subject for the Oldham Chronicle.

Sources of information are too numerous to mention in a short work of this kind, but the writer has accumulated a considerable store of colliery archives and other manuscripts as well as old printed books on mining. However, he is, as always, extremely grateful for the help given to him by Mr Eric Ridgway of the NCB, Keeper of Mining Record plans for the North West; Miss Glynis Matheson and her staff of John Rylands University Library, Manchester; the staff of Wigan Record Office and Wigan Reference Library and the staff of the Lancashire County Record Office.

Introduction

It was said many years ago that the true source of Britain's wealth is coal, and in spite of the discovery and exploitation of oil and gas from the seas around Britain during the last twenty years, that statement still contains a great element of truth.

It is not the purpose of this book to describe the origin of coal, a fossil fuel derived from the plant life that existed nearly 300 million years ago, and neither is it possible to include the many aspects of coal mining geology such as the sandstones, shales, mudstones and fire-clays associated with coal seams and the dislocations, intrusions, washouts and distortions of strata that make coal-mining in some areas so difficult. However, many books have been written on coal-mining geology and the reader is referred to these.

An eminent mining engineer stated in the 1890s that, 'a large modern colliery with its extensive and carefully elaborated equipment – including its various appliances for getting the coal and bringing it to the surface, for transmitting power through long distances underground, for causing great volumes of air to flow through miles of confined passages, for draining wide areas of underground workings and raising the water to the surface, for sorting the coal into various sizes and separating from it the intermingled stone and dirt, offers one of the most remarkable specimens of human activity and its triumph over matter.'

This 'triumph over matter' has brought with it during the past 150 years, a great catalogue of human suffering to those associated with it. A search through available records shows that upwards of 100,000 men have been killed in the pits during that period and very many more maimed for life or wracked by disease.

The harsh, unnatural conditions in which miners worked brought out their most manly qualities. During World War I, the Home Secretary said that he had been assured by the most eminent soldiers that miners carried to their work in the firing line the bravery and determination that actuated the industry throughout the country. Sir Douglas Haig said there were no more gallant or enduring men than the miners in France, and as a writer in *The Times* of that period put it: 'The miner as a rule makes an excellent fighting man. He is not of imposing stature, but he is tough and fearless. The pit has no mercy for the weedy or the timid. It calls for strength and stamina.'

Carlyle's words summed up the condition of the miner of the past: 'Venerable to me is the hard hand, venerable too is the rugged face. Is it not the face of a man living manlike? Hardly entreated brother! For us was thy back so bent! For us were thy straight limbs and fingers so deformed! Thou wert our conscript, on whom the lot fell, and fighting our battles thou wert so marred!'

Above: A well-constructed pit bank near Manchester in the last century. Note the 'kickup' tipplers, the fixed bar screens and the pump beam protruding from the engine house on the right of the photograph. A vertical winding engine worked a single cage in each shaft which ran on wooden guides

Below: Frog Lane Colliery, Coalpit Heath near Bristol, which was sunk in the mid-1850s and closed in 1949. The seams worked were Hard Vein, High Vein and Hollybush from two shafts 220yd deep. Note the ubiquitous saddle tank locomotive on the right. Manpower at closure – 258 underground and 79 on the surface

The Ownership of Coal

Since the Coal Act of 1938 became operative in 1942, all coal in Britain (apart from a few very small areas alienated by the Coal Commission during their short tenure of office from 1942 to 1947) belongs to the nation. The Commission sometimes alienated thin coal seams being worked in conjunction with a more important mineral such as fire-clay. Until 1942, in normal circumstances, the owner of the surface owned all minerals under his land with the exception of the 'Royal Mines' of gold and silver, and latterly oil. When the 1938 Coal Act came into force, the

valuation of the existing rights of the landowners in their coal seams was assessed at £66,450,000, and this was shared between them.

Unless an estate was very large, such as the Duke of Newcastle's estate in the Nottinghamshire coalfield (where the Wigan Coal & Iron Co leased 33,680 acres of coal from him at the turn of the century), the ownership of coal by landowners often mitigated against the successful and economic working of collieries. Surface boundaries have no relevance at all and surface rights only a minor relevance to the proper organisation of a colliery underground. Sometimes the areas worked from a pit were

An east Lancashire landsale colliery in the middle of the nineteenth century

Above: Powell Duffryn Steam Coal Co's Lower Duffryn Colliery at Mountain Ash. It employed close on 1,000 men during the 1920s and still employs 600. Note the beautiful valley beyond

Below: Cwmtillery Nos 1 and 2 pits at Abertillery, about 1908, which belonged to Lancaster's Steam Coal Collieries Ltd and employed 1,600 men in 1920

Horden Colliery, on the Durham coast between Sunderland and Hartlepool, employed 3,000 men after World War I and still has 2,000. The workings extend under the sea

not necessarily the areas which could most economically have been worked from that pit, but those for which the mine-owner had succeeded in obtaining leases. Landowners could not be forced to lease their coal if they decided not to do so. Amongst other points at issue were the barriers of coal left between properties, drainage systems, rights of support of the surface and way-leaves. Apart from the foregoing, much resentment was felt by the miners about the charge being levied upon the fruits of their labour for the advantage of individuals, who put neither work nor enterprise into a colliery concern.

As an example of the difficulties met with, the Head Surveyor of the Wigan Coal Corporation dealt with about 1,200 different lessors and their royalties before the passing of the 1938 Coal Act, and many large colliery firms were in a similar position. Nevertheless, many of the royalty owning families of this century had been colliery entrepreneurs in earlier times, such as the dukes of Norfolk, Hamilton and Devonshire, Earl Gower and many more. Some even remained colliery owners as well as royalty owners: the Earl of Crawford & Balcarres was not only chairman of the Wigan Coal & Iron Co and Wigan Coal Corporation right up to nationalisation, but also chief lessor of coal to the company; and the Earl of Ellesmere, successor to the famous Duke of Bridgewater, who only sold his rights in the Bridgewater Collieries in 1923.

In Durham and Northumberland, the Dean and Chapter of Durham and the Duke of Northumberland were prominent royalty owners, but there were many others like the Lambtons, Vane-Tempest-Stewarts, Millbankes, Delavals and Liddells. In Cumberland, the Lowthers and Curwens were prominent, in Flintshire the Mostyns, in Warwickshire the Newdigates, in Staffordshire Lord Dudley, and in South Wales the Marquess of Bute; these were nearly all at one time colliery proprietors as well as royalty owners.

The terms of mining leases and the system of royalty payments varied in different parts of the country, but one of the main provisions in colliery leases was the payment of a minimum or

Ashington Colliery at Morpeth in Northumberland early this century. A very large colliery employing 5–6,000 men between the wars. Note the squat ventilation furnace chimneys and the pumping engine house on the right

certain rent. This had to be paid whether any coal was worked in the area leased or not, and often it was tied to a certain tonnage. Certain rents could be anything from £1 to £5 per annum per statute acre of the whole area under lease. In many coalfields the royalty was paid on the tonnage raised, generally 6½d to 8½d per ton. In other areas it was from one-seventh to one-twelfth of the selling price per ton, whilst some coalfields paid a royalty on the acreage of coal worked. A variation of the latter system was the payment of a certain sum per foot thickness per acre, generally from £25 to £60. This system was common in Lancashire, except that the customary acre was the Cheshire acre of 10,240sq yd or 8yd to the pole. The royalties were in the region of £50 to £120 per foot thick per Cheshire acre depending on the quality and selling price of the coal.

The writer well remembers the visits of royalty surveyors every six months to

measure the thickness of the various seams in a royalty, to 'cast' the areas on the plan and calculate the acres, roods, perches and square yards, and finally the amount of money owing to the landlord. At the coal-face there was much arguing over thickness; the colliery surveyor always taking the landlord's surveyor to that part of the face where the seam was at its thinnest. After all this had been agreed, the colliery surveyors and the royalty surveyors adjourned to a local hostelry as the guests of the colliery company. It was the writer's job every Christmas to present one of the most important landlord's surveyors with a bottle of whisky. This particular surveyor was employed full-time by Lord Gerard, the principal royalty owner in the area, whose income from coal at one period amounted to £70,000 per year. Lesser estate owners employed firms of mining surveyors and engineers, of which there were many, to look after their interests.

About £6,000,000 per annum, or 6½d to 8½d per ton, was paid to private landlords in coal royalties between the wars, and in those days of depression and foreign competition, when a copper or two made all the difference between

Above: The Mary Pit, Lochore, Fife. Sunk by the Fife Coal Co, the first sod was cut by Mrs Mary Carlow, wife of the managing director, in 1902. The rectangular shaft was 28ft × 11ft and over 2,000ft deep to rich seams of navigation coal. It produced 250,000 tons annually at its peak, and pumped 90,000 gallons of water per hour. The headgear pulleys were made by Krupp of Essen

Below: Industrial panorama in east Manchester in 1950 showing Bradford Colliery in the middle of it. Owned by Manchester Collieries Ltd prior to nationalisation, difficulties were experienced with subsidence in the built-up areas under which it was worked

Above: A modern washery at Mosley Common Colliery, Lancashire, built by the NCB to deal with a large output. In some coalfields, large capacity washeries are required at many of the pits to deal with the inferior seams now being worked; equally, modern coal extracting machines load the whole seam onto the conveyor, including dirt bands

Below: Parkside Colliery, Lancashire, the only large, new colliery sunk in Lancashire since nationalisation, which has been designed and equipped on the very best principles of modern mining

failure and success in obtaining a contract, it was obviously not a light burden.

Many colliery proprietors amassed huge coal estates, an example being the Earl of Crawford & Balcarres at Wigan, who, at the beginning of the nineteenth century, bought up many estates adjoining his own considerable Haigh estate. His advice to his son and his grandson was to buy land which contained the most superior seams and then sell off the surface. His dictum was, 'Colliers we are and colliers we must ever remain'. During the middle of the last century, it was computed that his family owned sufficient coal to last 200 years at the not inconsiderable output produced by their pits at that time.

An example of a colliery owner acquiring a large mineral estate, with which the writer was intimately acquainted, was that of Col Henry B. H. Blundell of the Pemberton Collieries, near Wigan, who, before he sank his deep pits in 1870 at Pemberton, bought up many small estates and exchanged a considerable area of surface for minerals with Meyrick Bankes of the Winstanley estate who was himself a colliery owner. By that means he accumulated a mineral estate of more than 700 acres, containing over 40 million tons of coal, and held in addition valuable leases of coal from other adjoining estates.

There were many examples similar to the two quoted in every part of the country: the Fife Coal Co for instance possessed a vast freehold and leasehold mineral estate in the early part of the century, not to mention the Powell Duffryn Co in South Wales. Again, in South Wales, vast tracts of coal and other minerals in the Rhymney Valley were owned by the Rhymney Iron Co, in the Sirhowy Valley by the Tredegar Iron & Coal Co, in the Ebbw Valley by the Ebbw Vale Steel & Iron Co and in the Blaina Valley by the Nantyglo & Blaina Ironworks Co.

Some lands were held directly under the Crown, such as the New Forest, the Forest of Dean, parts of Derbyshire and the Duchy of Lancaster. The Crown also owned the minerals under the sea, the sea-shore and all navigable rivers. Any collieries wishing to work under the sea, or the other areas mentioned, had to make arrangements with HM Commissioners of Woods, Forests and Land Revenues, although in the Forest of Dean and parts of Derbyshire, there were ancient laws regulating the working of minerals, governed by special courts or bodies. An example of the latter was a friend of the writer's who, born in the Hundred of St Briavels in the Forest of Dean and having worked for a year and a day in the mines there, registered his claim as a 'Free miner' of the Forest, to work an area of coal there without payment of dues or royalties. These ancient rights were preserved by the Coal Act 1938.

The Colliery Owners

Up to the middle of the last century, the great majority of collieries were very small in terms of output and manpower. The average outputs per colliery for the main coalfields in 1854 were as follows:

Northumberland and Durham	68,500 tons
Cumberland	38,600 tons
South Wales	34,900 tons
Lancashire and Cheshire	27,200 tons
Yorkshire	26,300 tons
Scotland	20,600 tons
Derbyshire	19,600 tons
Staffordshire	14,500 tons

Britain was then the 'workshop' of the world, and to maintain its industrial supremacy there was an ever increasing demand for coal. There was also a great demand for export to foreign countries. The national output had risen from about 10 million tons per year in 1800, to 54 millions in 1850. During the next fifty years, there was another phenomenal increase to 225 millions in 1900. It continued to rise until its peak in 1913, when 287 million tons were raised. It took 3,024 mines, employing 1,107,000 men, to raise this vast quantity of coal. Between the two world wars, the British coal industry went through a very bad period, due to the world recession, with the consequent fall in the price of coal resulting in the closure of many pits, but more will be said about that later.

The table on page 17 gives the number of colliery companies and production units in each coalfield for the years 1898 and 1946, the latter being the year mines were nationalised.

The figures show a dramatic change in the industry during a working lifetime. From 1947, the year of nationalisation, up to 1980, the number of working mines has dropped from 1,296 to 219 and the number employed from 703,900 to 232,500, but due to complete mechanisation of the pits, the output has only decreased from 187½ millions in 1947 to 109⅓ million in 1980.

It will be seen from the number of companies quoted in the table that there was a great variety in the type of colliery owner, from a working collier who, along with two or three working partners, operated his own small pit, to a duke, marquis or earl who sat on the board of directors of a large colliery company. Some aristocrats owned their own collieries well into this century, examples being the Earl of Dudley in Warwickshire, the earls of Ellesmere and Lathom in Lancashire, the earls of Warwick and Waldegrave in Somerset, the Countess of Buckinghamshire in Fife and the Earl Fitzwilliam in Yorkshire.

There were many others who held the controlling interest in limited companies, such as the Earl of Crawford & Balcarres and his son, Lord Balniel, who were the main shareholders in the important Wigan Coal & Iron Co. The Queen Mother's family concern, John Bowes & Partners, owned nine collieries in Durham. Capt G. F. Bowes-Lyon was company secretary in the early 1920s. Their family seat in Durham was Streatlam Castle.

During the 1920s, the Duke of Sutherland, the Marquess of Londonderry, 14

Coalfield	1898		1946	
	Companies	Mines	Companies	Mines
Northumberland	80	112	38	79
Durham	91	246	55	152
Cumberland	25	39	11	16
Yorkshire	240	374	107	184
Derbyshire	88	161	39	83
Leicestershire	18	21	10	11
Nottingham	22	39	18	40
Warwickshire	18	25	15	16
Lancashire	151	327	42	94
Denbighshire and Flintshire	40	55	11	12
North Staffordshire	47	80	18	23
South Staffordshire and Worcestershire	213	244	33	47
Cheshire	13	21	3	3
Shropshire	38	70	10	14
Somersetshire	19	24	7	11
Gloucestershire	45	58	11	14
Kent	—	—	3	4
South Wales				
Monmouthshire	70	124	27	55
Glamorganshire	171	297	46	142
Breconshire	20	21	4	9
Carmarthenshire	42	45	7	32
Pembrokeshire	7	8	1	1
Scotland				
Clackmannanshire	3	6	2	11
Fife	33	52	12	35
Haddingtonshire	7	8	3	7
Lanarkshire	122	235	32	79
Dumfriesshire	3	6	1	4
Midlothian	14	19	11	23
Dumbartonshire	10	11	3	7
West Lothian	19	31	13	26
Renfrewshire	9	9	13	19
Stirlingshire	30	44	4	7
Ayrshire	34	95	11	38
Argyllshire	1	1	—	—
Sutherlandshire	1	1	1	1

earls, 5 viscounts, 16 barons and 47 baronets were directors of colliery companies. Some of these barons and baronets were colliery owners who had come by their titles in the New Year's Honours List, but there were many hereditary peers. Again, some had directorships of a number of colliery companies at the same time. For example, Sir David Richard Llewellyn of The Court, St Fagans, Glamorganshire, was a director of eighteen com-

panies. It must have been a full-time job just attending meetings. Sir David Llewellyn was the son of Alderman Rees Llewellyn of Aberdare. His son is the well-known Olympic Gold and Bronze medallist at show jumping and his grandsons, Dai & Roddy Llewellyn, are beloved by the gossip columnists. He took an active part in the management of seven of the eighteen collieries of which he was a director: he was chairman and chief engineer of Duffryn Aberdare Colliery Co Ltd, and chairman of five other companies.

The Viscountess Rhondda was on the boards of eleven colliery companies. She had been educated at Somerville College, Oxford University, sat in the House of Lords in her own right after a legal battle with the Lords, and was altogether a formidable lady. She was a member of the Court of Governors of the London School of Economics and

Above left: Sir Roger Bradshaigh, Bt, of Haigh Hall in Lancashire, a prominent early eighteenth-century colliery owner. Haigh was famous for its cannel coal which, apart from its qualities as a superb fuel, was carved into snuff boxes, candlesticks and other ornamental articles. His descendants continued to develop the collieries on the Haigh Estate, eventually forming the Wigan Coal & Iron Co in 1865 and the Wigan Coal Corporation in 1930. The 28th Earl of Crawford & Balcarres was Chairman of the Corporation at nationalisation and the last President of the Coal Owners' Association of Great Britain

President of the University College of South Wales. She wrote several books, and lived during the latter part of her life at Shere in Surrey and 70 Arlington House, St James's, S.W.1.

Lord Aberconway was a director of ten colliery companies, some of them with large pits in Yorkshire. He lived at 43 Belgrave Square, London, and Bodnant in Denbighshire. The 1st Lord Aberconway was an MA of Edinburgh

University, a Barrister at Law of Lincoln's Inn, a QC, the MP for Stafford, and author of *The Basic Industries of Great Britain*. Besides his colliery interests, he was chairman of John Brown & Co which built the *Queen Mary* and the *Queen Elizabeth*. His son, the 2nd Baron, was also a barrister and MP for Stafford. He was Chancellor of the Exchequer from 1908 to 1910 and succeeded his father as chairman of John Brown & Co and a director of colliery companies.

Viscount Furness, whose second wife, Thelma, was a friend of the Prince of Wales during the 1920s and '30s, was a director of five colliery companies in Durham, being deputy chairman of the Weardale Steel, Coal and Coke Co. He was also owner of the Furness Steamship Line and had interests in shipbuilding yards and engineering.

Sir A. F. Pease, Bt, MA, of the famous Durham Quaker family, was a director of five colliery companies in Durham. He had been educated at Trinity College, Cambridge, and was chairman and managing director of the family firm, Pease & Partners. He was President of the Mining Association of Great Britain in 1913–14.

There is no doubt that there were men of great intellect on the boards of some colliery companies who took an active and beneficial part in the management of them. It is equally true that there was a great number of colliery directors with little knowledge or ability, who were

Below left: Standing, from left to right: Col Swaine, Capt Holford, Lord Curzon, Lord Gerard, HRH Prince of Wales, Sir Kelly Kenny, Gen Brabazon. Seated, from left to right: Lady de Trafford, Hon Ethel Gerard, Lady Gerard, Christopher Sykes, Lady Randolph Churchill, Hon F. Gerard, Hon Mrs Oliphant. William Cansfield, Lord Gerard of Bryn, was one of the richest mineral owners in Britain at the turn of the century. He is seen standing fourth from the left between the Prince of Wales and Lord Curzon, Viceroy of India. Lady de Trafford was wife of another prominent royalty owner. The photograph was taken at Lord Gerard's residence, Garswood Hall, Lancashire, in 1898

incapable of advising on pit problems or men, and were content to drive up from the West End of London or some lush place in the Home Counties, put in an appearance at the director's meeting and draw their director's fee for doing so.

Of course nearly all management problems were left in the hands of the mining agent or general manager to the company, who generally had the backing of a consultant viewer. Many of these mining agents were extremely capable mining engineers who enjoyed considerable prestige in the mining community. They were paid excellent salaries, lived rent free in large houses and enjoyed a very high standard of living. The colliery agent to whom the writer was articled as an apprentice, superintended a very large colliery with coke and by-product works, brick and tile works, several farms and hundreds of houses. He lived for the last twenty-five years of his working life in two large and ancient manor houses with a large acreage of land attached to them. Both houses were rehabilitated and refurbished by the colliery staff and workmen, and it is doubtful if many top NCB executives at the present time enjoy a higher standard of living than this or many other former agents of the larger colliery concerns. The colliery had to make money, and as long as it did so, and paid consistently good dividends, the directors were happy to leave the agent alone. Emoluments came the way of these agents from other sources. Colin McLuckie, the inventor of a successful gas detector, told me of a friend of his who was agent to a colliery company in Yorkshire with five large collieries under his charge, who took on five articled apprentices each year at a premium of £500, giving him an income from that source alone of £2,500 a year.

The company for which the author worked may have been typical of many colliery companies. The chairman of directors was Major Cuthbert Leigh Blundell-Hollinshead-Blundell of Slaugham Place, Haywards Heath, Sussex. The other directors were his cousins, Major Stuart Upperton and Miss B. M. Upperton of Wimbledon Park, London, and his sister, Lady Lethbridge, of La Mancha Hall, Ormskirk, Lancashire. For generations the Blundells had followed the same pattern – Eton, Sandhurst and the Grenadier Guards. The management of the collieries was left to those who served them. The wife of one of the Blundells had been a Maid of Honour to Queen Victoria, the fine colliery village church having been built as a memorial to her.

After the 1926 strike, loss of exports and a decreasing home consumption led to cutthroat competition in the industry, resulting in the closure of many uneconomic pits and short-time working of those that remained. In an attempt to remedy the situation, the Coal Mines Reorganisation Commission was established in 1930 for the purpose of encouraging, and if necessary, enforcing amalgamations between colliery companies. The Coal Mines Act, passed in that year, had two parts: Part I required the establishment of central selling schemes and quota restrictions in each coalfield with powers over production and prices; the one set up in Lancashire, for example, known as Lancashire Associated Collieries, worked quite successfully. All contracts and sales had to be reported to the executive boards established to operate the schemes, and minimum prices were fixed for every class of coal. These executive boards consisted of coal-owners. Prices in markets supplied by a number of districts were co-ordinated. Each district was given a production quota and every pit allocated a share of this based on the standard tonnage of each pit. It was known for a colliery company to buy another out in order to obtain its quota.

Part II of the 1930 Act related to the amalgamation of colliery companies. The owners were, to a very large extent, against this and only a few large amalgamations were formed. Notable amongst these were: Powell Duffryn Ltd, an amalgamation of 14 companies with 43 collieries, employing 23,200 men; Manchester Collieries Ltd, an amalgamation of 7 companies with 13 collieries, employing 17,100 men; The Wigan Coal Corporation, formed from 4 companies employing 11,760 men at 12 collieries.

In Yorkshire 4 firms joined together to form Amalgamated Denaby Collieries Ltd, with 5 collieries employing 10,000 men. Another example was Doncaster Amalgamated Collieries Ltd, with 6 collieries and 15,600 men, which had formerly belonged to 5 firms. In Durham, Jas Joicey & Co and the Lambton & Hetton Co amalgamated to form Lambton, Hetton and Joicey Collieries, employing 12,000 men. In Northumberland 3 companies joined to form Hartley Main Collieries Ltd, employing 5,300 men.

In South Wales, Ocean Coal Co and United National Collieries formed Ocean and National Collieries Ltd with 12 collieries and 9,500 men. Another was Amalgamated Anthracite Collieries Ltd, a combination of 11 companies owning 22 collieries employing 8,000 men.

In Scotland there were already some large companies such as Wm Baird & Co of Glasgow and the Fife Coal Co, and very little more was attempted there.

The employers' organisation, the Mining Association of Great Britain, was formed in 1854; subjects engaging their attention being legislation, the settlement of principles determining the payment of wages, and the consideration of questions affecting conditions of employment, such as the limitation of hours of work, etc. The Association was a federation of district Coal-owners' Associations. Two Wigan men, Maskell William Peace and Sir Thomas R. Ratcliffe-Ellis, served more than fifty years between them in the position of Law Clerk and Secretary.

Eminent colliery owners and mining engineers from every coalfield took turns as president. From 1919 to the 1930s, however, Sir Evan Williams Bt, of Pontardulais near Swansea, was president. He was described to the writer by a colliery owner in his eighties, as 'a hard man' who led the colliery owners during the two disastrous strikes of 1921 and 1926 and the depression with its short-time working and unemployment for the miners of the early 1930s.

It was perhaps appropriate that the last president of the Mining Association of Great Britain was the 28th Earl of Crawford & Balcarres, chairman of Wigan Coal Corporation, whose ancestors had been mining coal and cannel at Wigan since the fourteenth century.

For many years the miners' unions had been pressing for public ownership of the pits, and the bad period they had been through since the end of World War I made them more determined than ever to achieve this. The election of a Labour Government in 1945 gave them the opportunity they needed: the Coal Industry Nationalisation Act became law at midnight on the 31 December 1946 when the pits and other assets of the colliery owners became vested in the Government. National Coal Board flags waved in the breeze at every pithead. Under the Nationalisation Act, a total sum of £164,660,000 had been fixed by national arbitration for the value of the colliery assets of the mining industry. This sum was divided by a national valuation board into district values; Scotland, for example, was awarded £21,837,000.

The National Coal Board had its first headquarters at Lansdowne House, Berkeley Square, London. The chairman was Lord Hyndley, supported by a deputy chairman and seven members. One was concerned with manpower and welfare, one with labour relations, two with production and the others with marketing, finance and science. There was a secretary, legal advisor, chief mining engineer, an advisor to the latter and a chief finance officer. Eight divisional boards were set up, each with a chairman, deputy chairman, labour director, marketing director, finance director, production director and secretary. Each division was divided up into areas under the charge of a general manager. The largest division was the Northern Division (Durham, Northumberland and Cumberland) with ten areas and general managers.

Alterations have taken place over the years, there now being five full-time members of the national board and five part-time members besides the chairman and the deputy chairman. Divisions have been abolished, there now being twelve areas. At each area headquarters, there is a director with a deputy or deputies, several production directors and departments for engineering, estates, finance, industrial relations, lawyers, marketing, medical service, mining engineering, purchasing and stores, public relations, road transport, scientific control and staff and statistics. Special projects, like the opening up of the Selby Coalfield, are separately staffed.

The chief difference between the old colliery owners and the NCB is that the former had to make the business pay and raise the finance for new projects themselves, whereas the Coal Board's constant losses, amounting to scores of millions of pounds, are written off by the Government and new developments are also funded by the Government.

Taking the premier Scottish firm, the Fife Coal Co as an example, during the sixty-four years from its inception to nationalisation in 1947, it never missed paying a dividend to its shareholders. During the depression years of the late 1920s and early 1930s, this was only 3-8 per cent a year. However, 10 per cent or more was paid in forty-eight years, the best years and their percentage dividends being 1891—37½, 1892—32½, 1900—50, 1907—40, 1908—30 and 1916—35.

It was without doubt a very healthy concern, due in some measure to being under very able management, beginning with Charles Carlow and followed by his son, Charles Augustus Carlow, gold medallist of Cowdenbeath Mining School and a certificated colliery manager. Other Fife Coal Co managers included Sir Charles Carlow Reid, another qualified mine manager, who became regional production director for Scotland in 1942 and Dr William Reid, who led the Technical Mission to the Ruhr Coalfield for the Government in 1945; and in 1952 became chairman of the Scottish Division of the National Coal Board. It might be said that with such profits, the Fife Coal Co could have done more for the welfare of its miners, but it was probably the most progressive firm in Scotland in that respect.

Mention must be made of a very important branch of the NCB, the Opencast Executive, which during the last forty years has been mining coal by opencast means and providing an example to other countries by the superb way in which many of its sites are restored after the coal has been won.

Working the Coal

Although there is some evidence that coal was used and presumably mined to a very small extent by the Romans in Britain – examples of this evidence being the coal ash found in the Roman lead smelting furnaces at Wilderspool near Warrington and the coal found in the recent excavations of Roman remains at Manchester – it was not until the thirteenth century that it was being mined at the outcrops in many of our exposed coalfields. The Domesday survey of 1086 does not mention the mining of coal, but it records iron ore and lead mining, so presumably no coal was being mined at that time. The term 'sea coal', applied to coal in many instances in early days, probably had its origin in the coal mined or washed out from the seams outcropping on the sea coasts, especially in Northumberland and on both sides of the Firth of Forth. A ship load of 'sea coal' is mentioned in a trial at Newcastle as early as 1269, and by the end of that century coal was produced on a small scale in nearly all the present coalfields of Britain except, of course, Kent. Coal was discovered in the latter coalfield in 1890 but it was 1912–13 before three collieries there produced any coal.

Coal was not popular as a domestic fuel, as the only habitations with chimneys were large stone buildings such as castles, monasteries and large manor houses. Certainly the huts of the labouring classes had no chimneys, the hearth being in the middle of the floor; the smoke had to find its way out through a hole in the roof. In this respect, wood or peat fires were preferable to coal fires. Even the wealthy had a prejudice against coal, as according to Galloway's *Annals of Coal Mining* published in 1904, Queen Eleanor cut short her stay in Nottingham in 1257 on account of the nuisance caused by smoke from coal fires.

However, by the end of the twelfth century, coal had become established in London as the fuel for lime-burning, brewing, smiths' work, smelting, dyeing, etc, but the smoke produced caused so much pollution and annoyance that demonstrations took place against its use, culminating in a Royal Proclamation ordering tradesmen to return to the use of wood and charcoal, the fuels they had used previously.

Nevertheless, the coal trade continued to grow: there is evidence that coal was used for lime-burning and for smithy fires in connection with the many castles, abbeys and churches built during the fourteenth century. The main sources of supply were the coalfields accessible to the sea such as Northumberland, but by the end of the century most of the inland coalfields were supplying, by means of pack horses, towns accessible by the primitive roads and tracks of the day. The introduction of iron grates and chimneys, for those who could afford them, made possible the increasing use of coal for domestic purposes. It was from the thirteenth century that mines were opened up in many inland parts of the country. Early records mentioned by Galloway include the following:

Northumberland and the Tyne basin The monks of Tynemouth Priory were the owners of extensive estates on the north bank of the Tyne which had an abundance of accessible coal. This was being worked before 1269, and by 1292 they derived an annual income of 61s 3d from their colliery at Tynemouth and 20s from their Wylam Colliery and a brewhouse there.

Durham A petition to Edward I in 1302 states, 'whereas it is lawful for every Freeman to take coal found in his own land', they were harassed by the bailiffs of the Bishop of Durham.

Nottinghamshire, South Yorkshire and Derbyshire coalfield It is obvious that coal was worked in this coalfield from early days, as Queen Eleanor could not stand the smoke of Nottingham in 1257, as we have seen above. A lawsuit regarding the 'right to dig coales' in Denby, Derbyshire, is recorded in 1306.

Shropshire The carting of coal is mentioned in a grant of a right of way in 1250 to the Abbot of Buildwas, and the digging of coal in the Clee Hills is referred to in a licence of 1260.

Forest of Dean In the time of Edward I (1272–1307), the mining laws of the Forest state that 'the sea cole mine is as free in all points as the oare mine'. The term 'sea coal' seems to have been in general use whether the coal was obtained from the sea shore or exported by sea, or neither. It may have been used to differentiate it from charcoal.

Staffordshire In 1282 William de Audlege, who held the manor of Tunstall, had a mine of sea coal worth 14s 8d a year.

Lancashire Coal was being obtained in 1294 from Colne by the monks of Bolton Abbey, and in 1295–6 Henry de Lacy, Earl of Lincoln, received a revenue of 10s from the production of sea coal at Trawden. Coal in Wigan is mentioned in deeds dated 1350, when Margaret Shuttlesworth exchanged land with Robert de Standish, but he reserved 'fyrstone (probably cannel) and sea cole if it be possible to find them in the lands mentioned'.

Wales Collieries were being worked at Mostyn in North Wales in 1294–5 according to Pennant, an eighteenth-century traveller and writer, and in South Wales a charter granted to Swansea in 1305 by William de Brews gave the townsmen licence to dig 'earth coal' for their own use, but not to sell to strangers. Coal was also worked in Monmouthshire in the reign of Edward I (1272–1307).

Cumberland No doubt coal was worked in the Whitehaven area from early times, but the available records of coal-mining only go back to the dissolution of the monastery of St Bees in 1553 when the estates were purchased by Sir Thomas Challoner. They were soon to pass to the Lowthers who were to dominate the coal-mining scene on the Cumberland coast until recent times.

Scotland The records of Holyrood and Newbattle abbeys furnish evidence that coal was being produced at Preston in East Lothian and Carridon in Linlithgow, on the south side of the Firth of Forth, as early as 1200.

Methods of working coal

Where a coal seam outcropped in the valley of a stream, or on the sea-shore or anywhere where the overburden had been eroded to expose the coal, it would be very easy to dig it out, and no doubt the very first supplies in ancient times would be obtained in this way. Following the seam down would mean removing the overburden by means of picks, wedges, pinch bars and spades, and carrying it away to tip in barrows or baskets. After this, the coal would be dug out and this would constitute quarrying or opencasting. The system would be economical with a fairly thick seam until the overburden became too thick, or until the rocks forming it became too hard for excavation with the tools mentioned above, as gunpowder was not used in British mines until the late seventeenth century.

A very early method of mining coal was to sink a series of 'Bell' or 'Beehive' pits. A number of these have been uncovered in most of the coalfields by the workings of the NCB Opencast Executive. The practice was to sink a shaft of 3–5ft diameter through the overlying strata, which generally varied from 7–20ft thick, and then work the coal all round the shaft until it became unsafe. Sometimes the shaft was widened, or 'belled' out, just above the seam. By this time, another pit or pits would be down to the coal, ready for opening up. It was a continuous process, the dirt from the excavation of a new shaft being used to fill up an adjacent disused shaft. The coal excavated formed a rough circle in plan, seldom more than 18ft in diameter and many being only about 12ft (see photograph above).

Generally these pits were 6–10yd apart. At Arley, near Wigan, a variation of this system was discovered in the middle of the nineteenth century when a tunnel was driven to divert the course of the River Douglas. This tunnel intersected workings near the outcrop of the Arley seam and found a series of octagonal chambers with vertical walls

Ancient 'bell' pit workings discovered by NCB Opencast Executive in Derbyshire

opening into each other by short and narrow passages, looking in plan like a honeycomb. They are marked on the 6in scale Geological Survey Map as 'Supposed Roman workings', but why this was assumed is not now known. They are certainly different to any workings ever found elsewhere. One thing that may have lent colour to the suggestion that they are Roman workings is the fact that Wigan was the site of the Roman fort Coccium, and these coal works would be easily accessible to the Roman road from Deva (Chester) to Luguvallium (Carlisle) and Hadrian's Wall. Be that as it may, an eighteenth-century mine plan of the area has the statement 'coal worked by the Danes, AD 1000'. One wonders how the workers of Bell pits coped with the accumulations of water in the disused pits adjacent to them, as the dirt excavated from a 5ft diameter shaft through, say, 12ft of surface cover to the coal would not fill more than one-third to one-fifth of a similar old Bell pit, depending on the thickness of the coal and the diameter of excavation in the coal.

A further development in the method of working (although it may have been carried on at the same time as Bell pits)

25

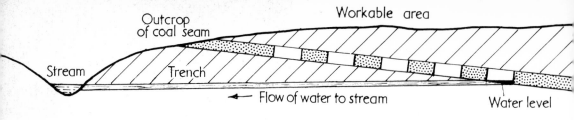

Fig 1

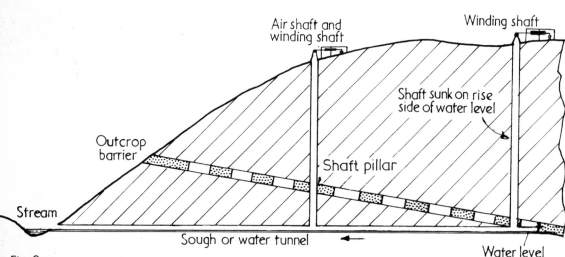

Fig 2

is the system in which roads were driven into a seam from a valley where both the land and the seam were rising away from the valley. The workings would be self-draining and, if the seam was thick enough, no unproductive work in moving dirt would be necessary. This is an ideal, cheap method of mining provided, of course, that the drift entrance was accessible to a convenient road by which the coal could be transported to customers. Where the coal dipped slightly into the valley sides, it was sometimes possible to cut open trenches and short tunnels to it from a lower level in the valley, so that a level road could be driven in the seam from a position lower than the crop and the coal then worked back to the crop (see Fig 1).

A drainage adit or sough constructed in the eighteenth century

The next development was the sough, or adit, or water-gate and shaft system. A sough or water tunnel was driven at a very slightly rising gradient from the lowest point in a valley, to intersect the seam or seams to be worked as far away from the crop as possible. From the point where the water tunnel entered into the seam, a waterway or level was driven in both directions, following the contour of the seam. By this means, all water met with between the water-level and the crop was drained away into the nearby surface streams or river. Shafts were sunk by the side of the sough and water level, both to facilitate ventilation and to extract debris or coal (see Fig 2).

It may be mentioned here that many famous adits or soughs were driven in various parts of the country during the seventeenth and eighteenth centuries, including those driven by Sir Roger Bradshaigh at Wigan in 1652–70, the Duke of Bridgewater at Worsley near Manchester from 1760, Sir James Lowther at Whitehaven and Sir Robert Cunningham at Stevenston in Ayrshire. The Duke of Bridgewater's soughs acted as underground canals to transport coal and eventually extended for 44 miles.

After the establishment of the drainage soughs and levels, shafts were sunk at intervals between the water-level shafts and the crop, and the whole area cut out into pillars. Up to the latter-part of the eighteenth century, it was the custom to leave pillars just strong enough to support the surface and many leases stipulated that pillars of sufficient strength for this purpose should be permanently left: 'the lessees must leave substantial and firm pillars of coal at proper due distances sufficient to support the roof of the hollows to prevent the upper grounds from falling in' – from a mid-eighteenth-century Lancashire lease. In many old workings seen by the writer, the pillars were on

The famous underground canals constructed by Francis, 3rd Duke of Bridgewater at Worsley near Manchester in the 1760s. They also acted as drainage soughs and were used as such until Mosley Common Colliery closed in 1969

Surveying underground last century, using a miner's dial or magnetic compass fixed on a tripod. The surveyor is either putting direction lines on a roadway or sighting a lamp on a similar tripod whilst carrying out a traverse survey of a coal-face

average 3yd square and the roads 2yd wide. Several variations on the system used to exist within the major coalfields:

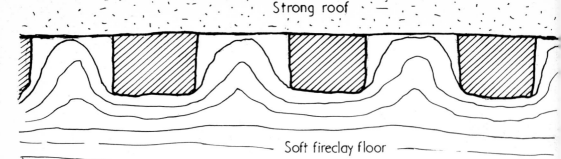

Fig 3

for example, a system of working was evolved in Lancashire which was partly pillar and stall and partly retreating longwall in panels.

Apart from having to contend with water and gas, many mines which were worked on the pillar and stall system, leaving small pillars, suffered from 'creep'. This was the bursting upwards of the floor of a mine with a normal fireclay floor due to the small pillars being pushed into the soft floor by the weight of the superincumbent strata. The writer has seen old roads completely filled up by 'floor lift' from this cause. At the beginning of the nineteenth century, John Buddle, the famous North Country Mining Engineer, devised a system of laying out the workings in panels, with panel barriers to prevent creep affecting the haulage roads and airways which ran between the panel barriers. Another less common danger

was 'thrust'. This was the crushing of small pillars when the floor of the seam was hard, and resulted in the fracturing of the roof, often letting down the surface or damaging seams lying above the one being worked.

The roadways in pillar and stall were driven by means of picks, hammers and wedges and pinch-bars. The seam was undercut or picked out to a depth of about 2ft 6in or 3ft, just above the floor, and sometimes a groove or 'nick' was cut with the pick down one side of the road. Then the coal was removed with the hammer and wedge, or blasted by gunpowder, or early this century by lime or hydraulic cartridges. Advantage was taken of the main natural cleavages or vertical breaks in the seam which run consistently in one direction, generally from NNW to SSE. Secondary cleats, not as well defined as the main cleats, are found at right angles to the latter.

CRUSH or THRUST

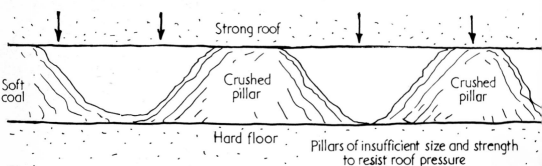

Fig 4

An underground roadway in urgent need of repair. Unfit for the passage of men

'Partings', or breaks parallel to the stratification of the seam, also facilitated the work of the collier.

In the 1860s John Warburton, a Lancashire colliery manager, noted: 'I have known men constitutionally old and finished at 34 years of age in consequence of working in straitwork [pillar and stall]. Besides the extra labour in this class of work there is a certain amount of oppressiveness owing to the small and confined space in which the work is performed. So confined is it, that the men not only breathe again the same air, but inhale a great amount of coal dust, so much so that their discharges are as black as coal. There is a great deal of hard labour in straitwork.'

Pillar and stall, highly mechanised, is the most common system of coal-mining

This roadway was 9ft high and 13ft wide when made about fourteen months before the photograph was taken. The height when the photograph was taken was 3½−4ft and the width 5−6ft

Colliers undercutting or 'holing' a fairly thin
seam at Coalpit Heath, Bristol, in about 1900.
The immediate shale roof above the coal is full of
'slips' and breaks and looks treacherous

Scene at a north country pit in the 1890s. It was
the custom in the Durham and Northumberland
coalfield for deputies to set the timber supports

in the USA, and during the 1940s and '50s some pits in this country adopted their methods and machinery, but this was later superseded by the latest methods of mechanised longwall. Unfortunately suitable conditions for mechanised pillar and stall mining in Britain are relatively few. Ideally the seam would be fairly thick, with good roof and floor and a relatively flat gradient. It would be fairly free from faults, firedamp (methane) and a liability to spontaneous combustion. The development roads should be able to stand well for considerable periods without ripping and this depends on the strength of the coal, and the depth and nature of the roof and floor. Given all this, the chances are that mechanisation properly applied to pillar and stall can result in a much higher output per manshift than other methods. However, most of the seams where this system could have been applied successfully in this country were worked out long ago.

The machines commonly used comprised shortwall or arcwall coal cutters for undercutting or overcutting the coal, drilling machines for drilling the shot holes over or under the cut or both, gathering arm loaders and shuttle cars for the thicker seams, or scraper chain conveyors instead of shuttle cars for the thinner ones. A shuttle car is a rubber-tyred electric truck with a capacity of 2½–6 tons which transports the coal from the loader at the face to a trunk belt conveyor or mine cars. It is fitted with a chain conveyor in the bottom to facilitate loading and unloading and is driven by storage batteries.

An alternative method is to use a Duckbill loader at the end of a shaker conveyor. This has a swivelling trough with a cast-steel shovel which digs into the loose coal and the shaking motion carries the coal along the troughs to a gathering conveyor which feeds the

Drilling a shot hole with a compressed-air drilling machine. Formerly this was done by hand machines

trunk belt or locomotive-drawn mine cars. Another and later method is to use a Joy, Goodman, or a Jeffrey Colmol Continuous miner at the face, which both breaks the coal from the solid and loads it into a shuttle car, or on to a scraper conveyor. This machine obviates the need for coal cutting, drilling and blasting.

Rotary percussive cutting machines were in use in pillar workings in Britain from the turn of the century, and performed a very useful service by cutting a working place 5–6ft deep and 15ft wide. Large numbers of these were used during the 1920s and '30s. The cut was made by the chopping action of a pick secured to the end of a steel rod which rotated as it made a reciprocating movement. The machine, mounted on a column, swept round in an arc, and on reaching the right or left extremity of the arc the cutting rod was fed forward by a handle and worm arrangement; up to six steel rods of increasing lengths were used. A team of two men could cut three places in a shift with one of these machines, made by the Hardypick and Siskol companies of Sheffield.

31

A successful cutter in 1903. Worked by compressed air, it could cut 7yd in 1 hour, with a depth of 4ft and a height of 4½in

Other machines used in pillar work were the arcwall and shortwall chain coal-cutting machines (mentioned above), developed from the longwall chain machine. The arcwall machine may be rail-mounted or on tractor crawlers. The cutting jib and chain is swung across the heading whilst cutting, and produces a horizontal cut at a selected horizon in the seam; the machine can be arranged to overcut or undercut. Some machines are designed to make vertical shear cuts which help to

Early disc coal-cutting machine. Photograph thought to have been taken at a Yorkshire Pit about 1900

bring the coal down with the minimum of blasting. Britain's first arcwall machine was installed in the Lambton Pit in County Durham in 1922. The shortwall machine acts in a similar manner to the longwall machine except that the jib is fixed in line with the machine. Cuts from 7–9ft deep can be made by both shortwall and arcwall cutting machines.

The longwall system of working coal is said to have originated in Shropshire. Nearly all the coal now mined in Britain is obtained by a very highly mechanised variety of this system. At the beginning of the century, although longwall was the chief method of working in Lancashire, Yorkshire, Staffordshire, Shropshire and Derbyshire, much coal was still produced by the pillar and stall method.

In the longwall method, the total extraction of the seam is commenced from the shaft support pillar, when, during the first decades of this century, the coal was removed in a long face, either in a straight or curved line or in steps or 'stepped buttocks', known as 'jows' in Lancashire. The line of the face was generally determined by the direction of the main cleat or cleavage of the seam, especially when the coal was 'hand got'. In most cases the face was set out parallel to the main cleat, although with stepped faces it was often advantageous to work with the cleat cutting across the steps at an angle. With soft coal, it was possible to remove the coal with a pick without first holing it, but generally the face was undercut or holed by hand or using compressed air or electric coal-cutting machines. The tons hewed per collier per shift on hand-got faces varied greatly. For example, 2½ tons was the average in the hard steam coals of Northumberland, but in the soft coke and gas coals of West Durham as much as 6 tons was common.

As the face advanced, roads were

formed by building 'packs' of stone 2–3yd wide on each side between the floor and the roof. The stone was usually obtained from the roof of the roads, thereby also increasing headroom, from the floor, or from fallen roof in the 'goaf' behind the face and between the packs of the roadways. At the beginning of this century, it was common practice for these drawing roads to longwall faces to be about 20yd apart. They were disected by cross roads every 60–100yd. The roofing down of these roads was often performed by contractors; the writer's grandfather had a considerable number of men employed on this work in the 1890s.

Although the first successful coal-cutting machines were introduced during the second half of the nineteenth century, they were not firmly established until the beginning of this century. Even in 1913, the peak year of British coal production, only 8 per cent of the output was machine cut. In early days the majority of coal-cutting machines were employed on longwall faces and were of the disc type. Cutter picks were attached to the periphery of the disc or cutting wheel, which was driven through gearing by a compressed air reciprocating engine, air turbine or electric motor. Its action was like that of a circular saw, but working horizontally, cutting a 5in-wide groove into the lower part of the seam. This type was superseded by machines having a revolving bar to which the cutter picks were fixed, until taken over in the 1920s by chain machines which acted like horizontal bandsaws.

Up to the 1940s and '50s, these chain machines were the standard machines used in the pits of Britain, and much good work was done with them. Credit should be given to American manufacturers who developed some very successful machines. The Scots were

Arthur Ashurst, master coal-cutter, who was Major C. D. Gullick's right-hand man when he had the coal-cutting contract at Pemberton Colliery, Lancashire

also pioneers in machine cutting, especially in thin seams. Cuts varied in depth but the average was 4ft–4ft 6in. The life of a cutter man could be hard and unpleasant. The works' magazine of Mavor & Coulson, manufacturers of coal-cutting machines, had this to say about the work of Arthur Ashurst, foreman coal cutter at Pemberton Colliery, Wigan, during the first thirty years of this century:

He was indefatigable in his work, beloved of the men who worked for him and by all the colliery officials. He represented that fine type of miner that is not sur-

The Denaby cutter/loader of the early 1950s. These machines cut and loaded to a depth of 4½ft. Dusty and not very reliable, they were used for about two years at a small number of mines

System of roof supports in the 1950s. Note fallen roof in waste area and the corner wall of pack. Conveyor jacks can also be seen

A coal plough at a Lancashire colliery in the 1960s. The face is supported by Continental-type roof supports – Schloms roof bars. The middle dirt, 24–30in thick, was loaded with the coal

Early double-drum Anderton shearer/loader working in the 1950s. The same type of machine, much improved, is the most common face machine in use today

passed in any industry; always 'mauling & scrawling' as he would say, but carrying on with that loyal perseverence in the fearless and cheerful manner so characteristic of the best type of British miner. He was a champion flitter with twenty or so men pulling by hemp rope a coal cutter which

had to be split up into several sections to get it from one end of the face to the other, some two miles, for in those days Arthur had a coal face just a mile long in semicircular form, with five machines following one another round. Arthur could personally drive a coal cutter with patience and that almost uncanny instinct of the experienced miner, beneath roof which he himself reported to be 'unfit for a cat to scrawl under', and shared in many accidents in his time.

Coal-cutting machines were continually improved during and after Arthur's time, the chief British manufacturers being Mavor & Coulson of Bridgeton, Glasgow, Anderson Boyes of Motherwell and British Jeffrey Diamond of Wakefield. Metallurgical problems which had caused trouble in the harsh conditions of mining were overcome, and the design and power of machines generally were improved.

The Anderson Boyes Meco Moore cutter loader was invented in the 1930s for seams more than 3ft 6in thick, but was much improved at Bolsover Colliery in the early 1940s. It was fitted with two normal jibs for cutting at the top and bottom of the seam, and a triangular-shaped jib for cutting the coal off at the back of the cut. The best results were obtained from this machine when the seam was more than 3ft 6in thick, the face 'on end' preferably with well-defined cleats in the coal, and the roof conditions good. Coal ploughs, long popular in Germany and Holland, have been and still are being used, and machines such as the Joy Longwall Buttock Miner and the Gloster Getter have been tried out, but the most popular machine is the Anderton Disc Shearer.

After nationalisation, when the Government made large amounts of capital available, great strides were made in mechanisation. Multi-jib coal cutters, which cut away most of the seam, mushroom, turret or capstan and spade jibs, and Huwood and other loaders were introduced.

The first conveyor was introduced underground in 1902 but their introduc-

Blaenant Colliery near Neath in Glamorgan, showing a Rackatrack which is the latest method of shearer haulage. The drift was put down in 1968 on the site of Cefn Coed Colliery to work the No 2 Rhondda seam. Annual output is about 335,000 tons with manpower of 670

tion at the more progressive pits was very slow until well after World War I. In 1931 only 21 per cent of the coal produced in Britain was delivered into tubs by conveyors, but by 1938 this figure had increased to 54 per cent, and in 1943 it was 66 per cent.

Great strides have taken place in the mechanisation of pits during the last fifty years. When the author first went down a pit in 1927, 56 per cent of all the coal produced at that colliery was still hand got; the rest was undercut by compressed air bar machines and heading machines, and after blasting was filled into tubs by hand. There were no conveyors. These small tubs (6 or 8cwt) were drawn to a main haulage shunt by ponies driven by young boys. There were 250 ponies. The tubs were drawn to the shaft sidings by either an endless rope, double-track haulage (with sets of empties going towards the face and sets of full tubs going in the opposite direction, lashed on to the rope with chains) or by a very fast main and tail rope haulage in trains of forty and sometimes sixty tubs. The shafts were over 600yd deep, of 16 and 18ft diameter and the 4 deck cages could hold 8 tubs or 32 men.

35

An underground main road junction at the turn of the century, showing an endless over rope haulage and ponies, probably used in the shaft sidings

A typical hand-got, hand-filled face of the 1920s. The photograph shows a tub of 8–10 cwt capacity, a collier's pick and hammer, and sprags against the top coal where the seam has been holed or undercut. A small jig or balance wheel is seen near the bottom right of the photograph. It is probably a balance jig where a bogie loaded with stone on a separate track draws up the empty to the face, the bogie itself being drawn up again by the full tub descending the incline

A large steam winding engine wound the cages up and down. The maximum output raised at any of the shafts was 2,000 tons in a day.

The large collieries today have coal faces equipped with disc shearers to break the coal off the face, rugged 'panzer' chain conveyors and hydraulic self-moving supports, producing, as far as the latest Advanced Technology Mining (ATM) and Heavy Duty Mechanisation faces are concerned, an average of 1,600 tonnes per day. The tendency today is for the coal to be conveyed to the surface by means of trunk belt conveyors of up to 5ft in width through drifts sometimes several miles in length, shafts being used only for ventilation, supplies and the raising and lowering of men. Where shafts have been sunk for coal winding during the past twenty years, they are often up to 30ft diameter and equipped with 20 ton skips. Some cages in man-riding shafts hold 100 men.

A four-deck cage at a well-constructed pit bottom

A radio-controlled shearer cutting coal at the face
at Kellingley Colliery, North Yorkshire Area –
the NCB's first 2 million tonnes a year mine

Perhaps a comparison of the statistics of a proposed ultra-modern deep mine with those of one of the most up-to-date mines of the time, sunk in 1868–70, may be of interest.

	1979 Proposed Park New Mine Staffordshire	1868 Pemberton Colliery, King & Queen Pits, Lancashire
Design Output	2.15 million tons per annum	½ million tons per annum
Daily Output	9,000 tons	2,000 tons
Life of Mine	50 years	100 years
Shafts	24ft diameter	18ft & 16ft diameter
Depth	approx 1,000yd	633yd
Total manpower	1,400 (including 70 staff)	1,800
Output per manshift	7.8 tons	1.1 tons
Markets	Power stations, general industrial and domestic	Export to Ireland and S. America; cotton mills, gasworks, railways, ironworks, engineering works, sugar refiners, domestic, etc
Estimated cost of mine	£140 million	£100,000
Seams	up to 10 workable seams	up to 14 workable seams
Reserves	300 million tons	50 million tons
Area	10 square miles	1½ square miles
		Note: The manpower was increased to 3,500 and the annual output to ¾ million tons in the years before World War I

Parkside Colliery, Lancashire. Photograph shows a cactus grab for loading the debris and the temporary lining of the shaft. The surface can be seen at bottom right

Shaft sidings at Frog Lane Colliery, Coalpit Heath, Bristol, at the turn of the century. The steps led to the haulage engine house. Note the steam pipe to the haulage engine and the pony, probably used for drawing empties from the back of the shaft to the haulage shunt. The shaft itself is probably behind the cameraman

Inset at Parkside Colliery, Lancashire, when sinking. The permanent concrete lining of the shaft can be clearly seen

Above: Coming to the surface at a Durham mine 60 years ago

Left: Rope-operated haulage for riding men forty years ago

Below: Photograph taken in 1929 of a junction on an over rope endless haulage road. The road is supported by steel joists set on wooden props; the cross-girder or 'running' girder at the junction and the junction itself is being supported by timber chocks

Above: Today, men are usually taken into the workings on special trains known as man-riding trains

Below: A surface endless chain haulage installation in which the chain slotted into a groove on the tub. Although these systems were simple and efficient to operate, they were vulnerable to the weather; snow in winter or heavy showers of rain caused track movement and could lead to loss of production for several days. This photograph was taken in the Burnley coalfield

Machinery for Pumping and Winding

Pumping

When the coal at a colliery above the level of the drainage sough or adit was nearing exhaustion, it became necessary in order to continue coal production to sink shafts below that level and either wind or pump the water up to the sough or to the surface. The shaft was always sunk several yards below the seam to be worked to form a sump or standage for water. If the 'make' of water was not excessive it was wound out by means of buckets or barrels. The barrels or tanks used for this purpose from the middle of the nineteenth century were equipped with valves for filling and emptying. The valve for emptying was made to work automatically by means of levers which engaged a trip wire in the headgear.

Other ancient methods of lifting water were the chain of buckets and the rag and chain pump. The former consisted of one or two endless chains hanging down the shaft (and going round a sprocket in the sump), worked by a sprocket drum generally activated by a water-wheel or horse gin at the surface, but sometimes by gangs of men. Buckets were suspended at intervals either from a single chain or from pins between two chains, the water being carried up the shaft in these buckets. Contemporary accounts complain about the large amount of spillage and also about the frequent breakages of the chain.

In the rag and chain pump, the rising side of the endless chain ran through a pipe in the shaft about 4–5in in diameter; secured at intervals to the chain and fastened round it were balls of sewn leather, stuffed with rags or horsehair and the same diameter as the pipe. These were drawn up the pipe by the chain, thus lifting the water up the shaft. Again, these pumps were generally driven by water-wheels or horse-gins.

Horse-driven gins, water-wheels and, in a few instances, windmills were also used to actuate bucket pumps (similar to the common well pump) in the shaft, through cranks, flat rods, 'L-legs' and pump rods down the shaft. The Laxey Wheel in the Isle of Man is a good example of this kind of arrangement.

The first type of steam pump was invented by Thomas Savery in 1698. It had no moving parts except the valves, and was described as intended for the 'Raising of water by the impellant force of fire', which Savery said would 'be of great use and advantage for Drayning Mines'. Unfortunately it was incapable of a total lift of more than 50ft and never came into general use. Sometimes two or more lifts were used, and it was not uncommon to sink a shaft close to, and half the depth of, the main pumping pit: the main pumps lifted the water to a drift, which communicated with the sump of the secondary pit, from where the pumps lifted it to the surface.

The first successful steam pumping engine was that invented by Thomas Newcomen in 1712. The bulk of the pumping in the coal mines of this country from the middle of the eighteenth century up to the latter part of the nineteenth century was done by the Newcomen-type atmospheric engine (much improved by Smeaton and others)

Engine house and headframe at the Duke pit, Hollinwood, Lancashire, in 1921. Note the pumping engine beam, the masonry pier supporting it and the shear legs seen end on

and Watt's development of it, the Cornish Engine. In early days, the pipes (known as pumps) were of wood, generally elm, but during the eighteenth century, cast-iron pipes were introduced.

John Farey, in his *General View of the Agriculture and Minerals of Derbyshire*, published in 1811, wrote:

In the early periods of mining a series of inclined wooden pumps, each of which was worked by a man, who sat and pulled up the bucket by means of a cross handle, were in use, called Churn Pumps, and also series of common Chain or Rag Pumps worked by a great number of men in succession in order to lift the water out of the bottoms of the mines; the last of which

were used on a great scale in Yatestoop Mine. Rag pumps were also used for collieries in some places and still are so at Whittington Moor. At Good Luck Mine, N.E. of Wirksworth, the Horse Gin was furnished with a cog wheel, pinion and cranks to work pumps, but is now disused. Windmills were on some occasions erected to work the water out of mines, of which remains are still visible N.W. of Monyash and at Dimple Mine in Matlock.

Waterwheels, working pumps by means of cranks and hence called cranks, are more common; at present Baslow Colliery, Dimsdale and Stanton Park Lane Quarries and Mines, the Gang Mines (where they have two waterwheels underground on Cromford Sough), Shallcross Colliery and Ticknall Lime Quarries are the only places where I saw such machines now in use, so generally have steam engines been adopted wherever great power of exertions are wanted.

I met with no Pumping Engine on Bolton [*sic*] & Watt's principle, at a Coal Pit; the old atmospheric engines, well contrived and executed, being thought to answer much better in such situations. Large steam engines for mines cost £2000 or more and consume 20 to 24 tons of coal per week.

On the 6 May 1725 Lord Harley, later 2nd Earl of Oxford, who had been on a visit to Lumley Castle, Durham, described an early atmospheric pumping engine thus:

From Chester-le-Street we go about half a mile to the left where there is a very large, fire engine for draining the coal pits there. The boiler holds eighty hogsheads. The fire stove consumes five fothers or sixty bushels of coal in 24 hours. The brass barrel or cylinder is 9 feet long. Its diameter 2ft 4in. Thickness of the brass 1½in. From the surface of the ground to the bottom of the water is 24 fathoms or forty eight yards. The water in the pit is 2 yards deep. From the surface of the water to the drift or level where the engine forces it out is 12 fathoms. It discharges 250 hogsheads in one hour; it strikes (as they term it) or makes a discharge fourteen times in one minute.

In the same place are two other engines for draining called Bob-gins and are moved by water turning a wheel. They all belong to Mr. Headworth, Dean of the Church and Mr. Allan. The weekly expense of those engines is £5, paid by the owners of the colliery to Mr. Potter the undertaker of the fire engine, the owners allowing whatever coals are expended.

As Farey stated in 1811, the old atmospheric engine was popular at collieries, partly because of the fact that there was then very little sale for slack, and therefore fuel consumption was not of prime importance, and partly because Newcomen's original design had been much improved by Smeaton, Brindley and other engineers. However, we must

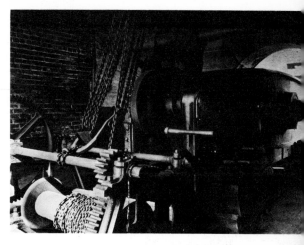

Cornish pumping engine known as 'The Duke' at Oak Colliery, Hollinwood, Lancashire. Built by Sir William Fairbairn, the famous Manchester engineer in 1846, the engine had a beam of 35ft long and a cylinder of 6ft diameter with 11ft stroke

The valve gear of the Duke Pit pumping engine as it appeared in 1921

look to James Watt as the inventor who, after 1776, brought the steam engine to be a really serviceable machine for pumping, winding and other commercial purposes. This was mainly through his invention of the separate condenser, whereby the steam, instead of being condensed in the cylinder, as in Newcomen's engine, was conveyed to a separate vessel where, by means of a jet

about these great pumping engines:

Upon the pulsations and vibrations of those ever-beating engines, which you can hear thumping and bumping all the night long, if you sleep near them, or rather lie awake through them — depend the dryness of the mine and the security of the miners. A peculiar noise these engines make at night. I sometimes fancied as I lay awake near one of them, that there must be some mysterious meaning in the sounds; first a heavy plunge down, bump and thump, then a short silence, then a kind of long drawn sigh, such as only a steam engine could draw, and then a rush of waters, enough to make you dream of drowning. Then in another moment, again bump, thump, sigh, cry, fuss and rush and all this without intermission.

The writer's father described the sounds made by the great Cornish pumping engine at Pemberton Collieries, Wigan, which worked from 1820 until 1917: 'At the beginning of every up-stroke, when the steam rushed into the cylinder, the engine made a loud noise like a groan, ending with a snort and at the end of every down-stroke there was a loud thud and the whole engine house shook. In the shaft itself there was a terrible clattering noise when the rubbing plates on the rods caught against the guide buntons.'

Often these pump rods were 14–16in square and each of them up to 60ft long. They were securely plated and bolted together at the joints. Generally there were at least two cast-iron pipe ranges in the shaft up to 2ft diameter; one from the bucket pump, normally at the bottom (sometimes there was more than one), and the other from the plunger or ram pump half way up. Connected with these were the great clack chambers. The buckets lifted the water to a tank from which the ram raised the water to the surface. The upward lift of the

Ancient valve gear of a late eighteenth-century colliery pumping engine by James Watt at a colliery at Hollinwood, Lancashire

of water, it became condensed and afterwards pumped out to be used as feed water for the boiler.

In the 1780s Watt introduced the system of cutting off the steam early in the stroke, the remainder of the stroke being effected by the steam's expansive action and this was the basis of the single-acting 'Cornish' pumping engine. Generally these engines worked from 6 to 10 strokes a minute, the engineman working the tappets by hand for the first 5 or 6 strokes and then connecting them to the plug rods.

Leifchild, in his *Our Coal and Coal Pits*, written in 1856, has this to say

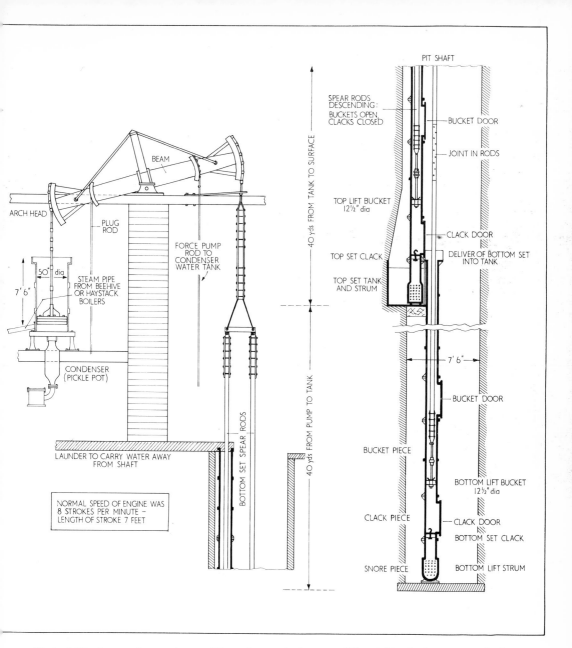

Clarkes' No 1 pumping engine at Winstanley, Lancashire

bucket and the downward thrust of the ram equalised to some extent the work of the engine.

There were buntons at frequent intervals acting as supports for the pipe ranges and guides for the pump rods, so when the shaftmen rode down in their little tub, there were many obstacles in their way. The job of pitman or shaftman in these old engine pits was a very dangerous and unpleasant occupation. There was much wear and tear, and whether it was a case of changing a broken spear rod, a clack or a bucket, or packing the ram, it was all heavy work performed with the aid of the capstan engine at the surface. For instance, at a pumping pit known to the author, it took 8 hours to pack the ram.

These pits being downcasts, besides being wet and full of equipment, were bitterly cold. Sometimes inspections had to be made 'riding the Loop' − a chain sling at the end of the rope. Also it would be necessary sometimes to ride escort on a massive baulk or other heavy equipment. Needless to say, there were many accidents and also many miraculous escapes in these old shafts. The writer remembers three shaftmen falling down a pumping pit when he was a boy. One hit a baulk and bounced into a mouthing, another was caught on an obstruction. Both of these men were unhurt, but the third was killed. Working long hours under such conditions resulted often in a short working life.

From the latter part of the nineteenth century, there were great improvements in the pumping equipment of collieries. Horizontal instead of vertical engines became common, working the pump rods in the shaft by means of flat rods and L-legs; then complete, self-contained pumping engines, worked by steam or compressed air, were placed in pump houses near the bottom of the shaft. Early this century electric three-throw ram pumps were introduced, both as inbye pumps and main pumps; many of the latter have since been replaced by centrifugal, or turbine pumps, which, much improved, are used today as the main pumps at most collieries.

Winding

In early times, and at shallow shafts, coal was sometimes carried up ladders in baskets, often by women. This was particularly true in working the Edge coal seams of the Lothian coalfield in Scotland. Otherwise the winch, crab or jack roll was used, generally fitted with two handles.

With the sinking of deeper pits, more effective methods of winding were required, the cog and rung gin being one of the first inventions. The disadvantage of this was that the horizontal drum was fixed over the shaft itself; this was operated by horses through a horse arm secured to a vertical shaft on which a large cog wheel was fixed, which meshed with a pinion on the end of the drum shaft.

A great improvement, and the most common method of winding throughout the eighteenth and first decade of the nineteenth centuries, was the whim gin or whimsey. This consisted of a large-diameter drum mounted on a vertical axle or shaft, turned by horses hitched to a long wooden arm or lever secured to the axle. The ropes from the drum went over two pulleys on the top of a head-frame over the shaft, and thus into the shaft. As the pits became deeper, more and more horses were required for the winding gins to produce the required outputs, and thus increased considerably the cost of winding (see Fig 6).

Other methods were the water balance and the water-wheel. The water balance consisted of a 'C' wheel on a shaft equipped with a brake, with the rope lapped 2½ times round the 'C' wheel. A tank of water descending, being heavier than a cage containing a tub of coal, brought the latter to the surface, whereupon the brake was held. The tank was emptied of water, and then brought up the pit by the descending cage and empty tub; the tank again being filled with water and the process repeated. If there was a drainage sough from the bottom of the shaft (as at the Duke of Bridgewater's Worsley Colliery where the system was used), it was ideal, but if this was not the case, then the water had to be pumped again to the surface (see Fig 7).

The water-wheel, with a double set of buckets and a movable water trough to

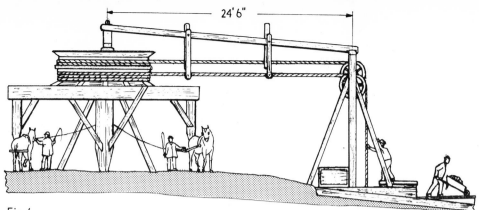

Fig 6

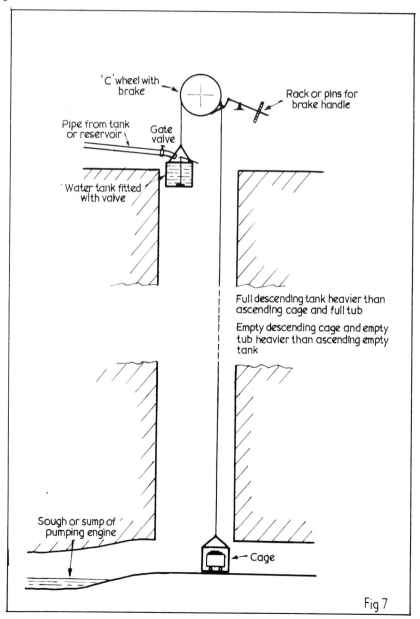

'C' wheel with brake

Rack or pins for brake handle

Pipe from tank or reservoir

Gate valve

Water tank fitted with valve

Full descending tank heavier than ascending cage and full tub

Empty descending cage and empty tub heavier than ascending empty tank

Sough or sump of pumping engine

Cage

Fig 7

The valve gear of the large double tandem compound winding engine at Astley Green Colliery, Lancashire, erected in 1913 by Foster, Yates & Thom of Blackburn. Its specification was: High-pressure cylinder, 35in diameter; low-pressure cylinder, 68in diameter; winding drum, bi-cylindro-conical, 17ft 2in to 27ft 0in; weight of drum and shaft, 99 tons; hp of engine, 3,000; static load on winding rope when winding coal, 38 tons; depth of wind, 872yd

set it going in reverse for winding up or down, was used either where there was a convenient stream of water or by means of the water pumped up the pit by a Newcomen engine.

Towards the end of the eighteenth century, however, the steam engine had been greatly improved, and although Boulton and Watt had certain patent rights up to 1800, there were many so-called 'pirated' versions of their engine.

J. Aikins, in *A description of the country from 30 to 40 miles round Manchester*, published in 1795, states that Messrs Bateman & Sherratt of Salford, had 'improved upon and brought the steam engine to perfection. They are now used in cotton mills, and for every purpose of the water wheel, where a stream is not to be got, and for winding up coals from a great depth in the coal pits, which is performed with a quickness and ease not to be conceived.'

John Farey wrote of the Derbyshire area in 1811:

Within a few years past, a new use of steam engines has been made by applying them to the winding or drawing up coals and minerals; these small steam winding engines are called 'Wimseys', the first of which in these districts, was erected at

Steam compound haulage engine in overhead engine house in the High Vein seam near the shaft at Frog Lane Colliery, Coalpit Heath, Bristol, in about 1900. The high pressure cylinder was of 15in diameter and the low pressure cylinder of 26in diameter. The condenser can be seen in the right foreground. This engine worked a plane 2,000yd long

Oakthorpe Colliery in Measham in 1790, as I am informed; they are now so common at the larger collieries, some of which have three and even four in their field, that I have noticed more than 50 of these wimseys in Derbyshire and Nottinghamshire. The cost of these, complete for drawing at one pit, may be about £500 each and they will raise 5 to 12 cwt. of coals at a time.

These were beam engines, but before the middle of the century, vertical winding engines became common. The cylinder or cylinders stood vertically on the ground floor of the engine house, and the rope drum (or reels if flat ropes were used) was placed on stone pillars above the cylinders, the drum shaft crank or cranks being joined by the connecting rods to the piston rods. At the same time the horizontal engine was being developed. At first, these were equipped with one cylinder and hand-operated valve gear, but during the latter half of the nineteenth century, when there was a great expansion in the coal industry and many deep pits were sunk, the horizontal winding engine with two cylinders was brought to a high state of perfection. Some of these were very powerful with 40in diameter cylinders and large rope drums. Many

49

drums were 20ft or more in diameter if horizontal, and over 30ft if conical or cylindro-conical.

The ropes in use on winding engines were at first round hemp ropes on the horse-gins, then flat hemp ropes on the first steam whimseys; flat, iron wire ropes came into use at pits from about 1840, and round, iron wire ropes soon followed. For many years now, special steel wire ropes have been used.

As a boy, the writer was taken by his father into a very imposing cathedral-like engine house; it was 116ft long, 38ft wide and 55ft high, containing two magnificent winding engines. They were an awe-inspiring sight, the great rope drums dominating the scene. One was a scroll drum, 30ft 10in in diameter, and the other, a 20ft diameter cylindrical drum. These great steam engines were typical of the many installed at the deeper pits in every coalfield during the last half of the nineteenth and the first quarter of the twentieth centuries. It was fascinating to hear the sharp clanging of the signal bells and then to see the engine set in motion, the sweep of the huge cranks and connecting rods and the almost poetic motion of the eccentrics, link motion and valve-operating gear. It very quickly gained speed in a completely effortless manner, the cage in the shaft soon attaining a speed of a mile a minute, but no sooner had this been achieved than the engine was thrown in reverse which quickly reduced the speed until, by a gentle operation of the brake, the ascending cage was brought level with the pit bank.

All the bright parts of the engine were burnished until the iron and steel shone like silver. The engine houses were lovely places to go during the winter time; the warmth and the smell of steam and oil, once experienced, are never forgotten. Many of these steam winding engines have been replaced in recent years by electrically driven ones, and all new NCB pits are equipped with electric winders.

Other large engines at a colliery comprised air compressors, which came into general use towards the end of the nineteenth century, fan engines for driving the main ventilation fan, introduced at many pits from 1870 onwards, and hauling engines situated on the surface to drive rope haulages underground by means of ropes going down the shafts, sometimes encased in wooden pipes.

For all these engines, steam was raised in various kinds of boilers. One of the earliest types was the beehive boiler. Another early type was the wagon boiler, almost the shape of a 'covered wagon'. By the early part of the nineteenth century, egg-ended cylindrical boilers, about 5ft diameter and up to 35ft long, had come into use. At the same period, Trevithick invented the Cornish boiler with a single flue running through it near the bottom. By the 1840s, Fairbairn of Manchester had brought out the Lancashire boiler, similar to the Cornish but larger in diameter and with two flues. Over the years this was improved and became the best type for use where there was an intermittent load. Early this century, however, at some of the new collieries, Babcock & Wilcox or Stirling water-tube boilers were installed.

Some of the larger collieries, especially those with coke plants, achieved a high degree of self-sufficiency during the period before the electricity supply grid system became widespread. The author served his apprenticeship at a colliery where, in 1930, the exhaust steam from the winding, hauling and fan engines went into steam accumulators, and these, with a certain amount of steam from the boilers, drove the mixed pressure turbo air compressors. All the electricity used in the colliery was pro-

50

A modern electric winder at Parkside Colliery, Lancashire, with four ropes to each cage

duced by generators worked by gas engines, fed with waste gas from the coke ovens. This waste gas was also used to raise steam in a battery of boilers at the coke plant.

Few books about miners mention the men who had the very important task of installing and maintaining all the engines and machinery on a colliery – in some cases they built their own stationary engines, locomotives and

Hem Heath Colliery at Trentham near Stoke-on-Trent originally consisted of a single shaft of 820yd deep when sunk in 1924–6, connected to Stafford Colliery. This photograph shows No 2 shaft sunk by the NCB between 1950–5 which is 1137yd deep, 24ft in diameter and concrete lined. 'A' type headgear with winding from two insets by two 2500hp electric winders. The cages are 3-deck and run in rigid steel guides. Two radial flow fans with 1500hp motors

wagons. Large colliery companies owned thousands of railway wagons which they built and repaired, maintaining expert staff to carry out this work. Carpenters, blacksmiths and bricklayers carried out all maintenance and repair work in the shafts, and on occasions were faced with absolutely daunting tasks after a shaft accident, sometimes with both cages, ropes and girders at the bottom of the shaft and with some of the buntons broken off. It is not often appreciated by outsiders that the craftsmen at a colliery covered almost every trade – joiners, carpenters, blacksmiths, fitters, tinsmiths, saddlers, wagon builders, tub builders, pattern makers, bricksetters, slaters, painters and glaziers, electricians, etc. In addition, there were often men with various kinds of expertise, such as drainers, horsemen and farmers attached to the colliery farms, and also platelayers, etc. Engineering works were established in the main centres of every coalfield to supply winding engines, compressors, pumps, ventilating fans, screens and every other type of equipment.

The Distribution of Coal

The classification of coal

Ignoring lignite, which is the connecting link between peat and bituminous coal, the chief types of coals are gaseous, semi-bituminous and anthracite, but there is a gradation of one form into another, so the dividing line can be rather blurred.

Cannel is the best of the gas coals, giving off large quantities of gas, producing a good coke and a large percentage of valuable by-products. There is now very little cannel left in the country, but since the advent of North Sea natural gas, coals of that class are no longer in great demand.

Bituminous Coals vary in composition and are used for steam-raising purposes, as house coal and for making coke; they were also used for gas production in former days.

Semi-bituminous coals are free-burning, non-caking coals, of great value for steam-raising, metallurgical and domestic purposes. At the turn of the century, the better dry, or smokeless, varieties, mostly produced in South Wales, were in great demand for ocean-going liners and warships.

Anthracite consists almost entirely of carbon; it is non-caking, does not ignite easily, burns with a feeble flame, but gives off intense heat. It is used for domestic central heating boilers and stoves, for smelting and other metallurgical uses, for heating kilns and all other purposes where a steady smokeless heat is required. Again, anthracite is produced in South Wales.

Preparation of coal

In the last century only the best seams were worked; hence, there was very little dirt in the seam, and very often the collier riddled it underground and stowed the slack. If he was careless and filled his tub or basket with any dirt, he was either fined, or he was refused payment for the tub. Picking belts – slow-moving belts, conveying coal to the wagon, from which dirt is picked out – were introduced about 1870, but screens had already been in use for many years, especially in the North East. Washeries, generally of the trough type, were also introduced about 1870, mainly to clean slack for use in coke ovens.

During the last century there have been tremendous strides in the preparation of coal for the market, by screening and washing. Under modern systems of mining, the whole seam, including dirt bands, is broken from the face and loaded on to the face conveyors. This means that the mineral wound out of the pit often contains only 60–80 per cent coal, and highly efficient washeries with a very large capacity are required. It also means that large quantities of dirt have to be conveyed from the washery to the tip daily.

Transport

The coal was generally loaded into wagons under the screens, the wagons being coupled together to form a train and weighed on a weighbridge. Weigh clerks and brakesmen became adept at this, the brakesmen controlling the wagon by using a heavy stick on the brake handle, without ever actually stopping the wagon. One clerk read and shouted out the tare, another the gross weight, whilst a third did the booking.

Wagon inspectors saw that every wagon was in good order before loading. The following instructions were given to brakesmen and wagon lowerers at a large colliery in 1920:

You are hereby appointed a person to lower wagons. Before setting any wagon into motion it is your duty to give warning to all persons employed on or about the screens and to assure yourself by personal inspection that every person is out of the way and that all is clear. When coupling or uncoupling wagons in motion the coupling pole must be used. Before loading any wagon, you must see that it is thoroughly clean and the wagon doors secure. You must not push at any wagon buffers.

Nowadays the NCB own relatively few wagons, but before nationalisation there were close on 600,000 privately owned coal wagons, most of them owned by the colliery companies. The larger colliery companies built and repaired their own wagons, which had to conform to a standard specification issued by the Railway Clearing House. From the middle of the nineteenth century, mineral wagons had developed from the 6 tonner to the 8 tonner and then after 1887, when no more dead- or dumb-buffered wagons were allowed to be built, to the 10 ton wagon in the early days of this century. A new standard specification was issued in 1923, and from 1 July 1924, no wagon smaller than a 12 tonner was to be built.

Part of the duties of the writer's father was responsibility for the building, and maintenance in good running order of close on 2,000 mineral wagons. He had a large, well-equipped wagon shop, and a wagon paint shop. The staff consisted of 8 wagon builders and 4 apprentices, 2 blacksmiths on wagon work only, 2 wagon shop labourers, 2 wagon painters and 3 wagon inspectors. The solebars, headstocks and other parts of the underframes were made from American oak, and when that was unobtainable during World War I, Australian jarrah was used. As an apprentice of nineteen, my father could build a complete 10 ton wagon in a week, which was the record at that colliery. For that he was paid the princely piece rate of £2 18s 0d. Nowadays all mineral wagons are constructed of steel, of 16, 20 or 30 tons capacity, equipped with side and/or bottom doors.

Formerly many wagons used for unloading at ports and for tipping into canal barges, were fitted with end doors, (in addition to side and bottom doors) the wagon being up-ended, causing the coal to slide out into the hold. A variation was three boxes on a wagon, the boxes being lifted off the wagon bottom by special dock cranes and emptied into ships' holds by opening bottom doors; this type was used for export orders from Liverpool last century and early this century. These were particularly good for distributing the load evenly in the hold, and the shorter drop reduced the likelihood of large coal breaking up. By the 1930s, rotary tipplers had been introduced at public utility and other industrial concerns by means of which loaded wagons were lifted over and emptied without the use of any doors at all.

Today much of the coal to power stations and large industrial customers is transported in 'merry-go-round' trains, the doors in the base of the special wagons being automatically opened, allowing the coal to drop into large underground hoppers, without uncoupling the wagons. The coal is fed from the hopper on to a conveyor belt that carries it into the stocking ground

Two 'Austerity' saddle tanks moving a heavy coal train at Bickershaw Colliery, Lancashire

or the power station. The loading of these merry-go-round trains is rapid at most collieries, a modern loading bunker being capable of filling a 30 wagon train with 1,000 tonnes of coal in 30 minutes.

Locomotives are required at collieries to shunt and marshall wagons; earlier this century there were thousands of 0−4−0 and 0−6−0 saddle and side tanks at work in the coalfields as well as larger engines. Some of these were built by the larger colliery companies in their own workshops: Wigan Coal & Iron Co, which owned upwards of thirty locos at their peak, built a large proportion of these themselves. Sidings at a large colliery were very extensive, there being 17 miles of them where the writer served his apprenticeship. A staff of plate-layers, under a main line trained foreman, was kept constantly busy. Besides sidings, there were hundreds of miles of private mineral lines in the coalfields, connecting with main lines and canals.

The wagons were placed by the colliery locomotives in sidings from where they could be picked up by main line railway company locomotives and hauled to marshalling yards to join hundreds or thousands of others. They were then sorted and made up into train loads according to destination. This was a colossal operation, and at the largest yards, the wagons would be pushed over a 'hump' and allowed to run by gravity to their appointed road. As many as 10 per minute could be pushed over the hump, spreading out fanwise into different roads. The wagons, when emptied at their destinations, were marshalled according to ownership and returned to their respective collieries. This led to a tremendous amount of extra shunting, with its attendant overheads, which caused the pooling of wagons to be advocated in the 1930s, but it never came about.

At the ports, many different methods

'Oceans' of coal at Aintree sidings, Liverpool, on 26 April 1911. Colliery wagons were marshalled here from all over Lancashire, Yorkshire and parts of the Midlands, their contents for export to Ireland, South America and elsewhere and to supply the many industries in the Merseyside area

were employed for unloading wagons, dependent on whether the fuel was small coal or large. In the case of the former, speed was the only consideration, but with the latter, anti-breakage devices were of paramount importance and loading was consequenly slower.

Markets

Before World War I, one quarter of the total output of Britain (more than 50 million tons) was exported, most of it to European and Mediterranean countries. The biggest customers, who took almost 30 million tons between them, were Germany, France and Italy; considerable tonnages also went to Russia, Sweden, Denmark, the Netherlands, Spain, Belgium and Egypt. Argentina, Uruguay and Brazil took quite large tonnages, especially Argentina which bought 3,365,000 tons in 1912.

British coaling stations spanned the world. Great liners proceeding from European ports to the Orient, whether British or foreign, could not reach their destinations without British coal from the maritime coaling stations at Gibraltar, Port Said, Aden, Colombo, Singapore and Hong Kong or Cape Town and other South African ports if they went by the alternative route. There was hardly an area in the world that was not within the reach of a British coaling station. These were kept well stocked by a great fleet of colliers.

London was formerly a very important consumer of coal. As regards house coal alone, one has only to look at the great stacks of chimneys on the large Georgian, Victorian and Edwardian houses to realise the extent of the trade. Until the advent of the canal and railway, nearly all the coal came by sea from the Tyne and Wear ports. Famous firms such as Charringtons, Corys and Bowaters spring to mind in connection with this and other aspects of the coal distributive trade, especially during the nineteenth century from South Wales.

Coke has been made in increasing quantities, in beehive and square ovens, since the early eighteenth century, and since the late nineteenth century in patent ovens which provided for the extraction of by-products. A considerable quantity of the coke produced was used in blast furnaces for the production of pig iron. Until the 1930s there were many coke plants associated with collieries in Britain. The National Coal Board produced approximately 4½ million tons of coke in 1979–80, slightly more than ½ million tons being exported. Patent smokeless fuels are produced and also chemical and secondary by-products.

The products that can be derived from coal are legion: tar, dyes, scents, explosives, motor spirit, pitch, ammonia, rubber, margarine, light and heavy oils, carbolic acid, plaster of Paris, smelling salts, sal ammoniac, sulphuric acid, saccharin and even sweets made of vanilla or essence of almonds.

The table below gives the percentages of the chief inland uses of coal at the beginning of this century and in 1980.

	1905	1980
Railways	8.29	—
Collieries	7.22	—
Coasting steamers	4.81	—
Factories	24.07	
Chemical works, potteries and glass works	3.61	20.5
Domestic	21.28	
Steel and Malleable ironworks	7.95	
Blast furnaces	11.64	9.5
Other metallurgical works	1.30	
Gasworks	9.83	—
Power Stations	—	70
Percentage of total output exported	28.44	2
Total raisings (million tons)	236	123.3

Accidents, Legislation
and Diseases ·

When the author started his apprenticeship in the survey department of a large colliery some 50 years ago, the average number of men killed in the pits year after year always seemed to be around 1,000, and another 130,000 were injured each year, some of them seriously.

Fatal accidents reached their peak in 1910, when 1,818 persons lost their lives. By 1938 the number had dropped to 858, and in 1947, the year the mines were nationalised, to 618. Thereafter the number of fatal accidents decreased to 420 in 1952, 216 in 1965, 91 in 1970 (the first time it had dropped below 100 in centuries) and 30 in 1979–80. Of course the number employed at coal-mines has also dropped, from 1,027,539 in 1910 to 233,200 in 1979–80, but as can be seen, the incidence of fatal accidents has dropped very dramatically. This is due to the great improvements in mining techniques and safety measures since World War I, but especially since the nationalisation of the mines.

Since 1850, when systematic records were first kept, over 100,000 men and boys and a small number of women have been killed at British coal-mines. Falls of roof and sides accounted for, very approximately, half of these deaths. Seen beside a statistic, the term 'fall of ground' conveys nothing at all of the terrible agony suffered by those who died from this cause. In some cases death was sudden, but in many others it was a slow excruciatingly painful death. Many of the deaths on the haulage roads and in shafts were just as terrible.

As a teenager, the author was extremely shocked when a deputy he had been with all morning was killed on his way out of his district, his body mangled by a train of 40 tubs on what was said to be the fastest 'main and tail' haulage in Lancashire. His name was Joe Gaskell, and it was a moving experience riding up the shaft on the following Saturday when the men sang the minstrel song 'Poor Old Joe'. The loud thumps echoing in the pit shaft as the cage slipper blocks passed over the joints in the railway line-type guides sounded like the beat of a drum. The author knew men who fell down shafts in three separate accidents. They were all experienced shaft men but they all neglected to take the precaution of wearing the safety belt provided. In former times, over-winding the cage or basket frequently caused casualties.

Only accidents in which there was great loss of life came to the attention of the public at large. The great majority of these were caused by explosions of methane and coal dust, the latter especially having devastating effects. At very odd intervals there were different kinds of disasters, such as that at Hartley Pit in Northumberland in 1862, where 204 people died from suffocation when they were entombed in the pit, and the inrushes of water at Redding in Stirlingshire in 1923 and at Montague Colliery in Northumberland in 1925, when a total of 78 lost their lives at both collieries.

The writer learnt the story of the Hartley disaster, when a small boy, from his grandfather. His father was under-manager at the colliery at that period, and I used to be shown a drawing of the great beam engine, the broken beam of which caused the calamity by falling down the single shaft, demolishing the partition in the shaft, and bringing down much of the shaft walling, thus blocking the shaft. It does not require much imagination to perceive the effect on a mining village of such an accident, wiping out a large number of the boys and men of working age, leaving only the old and very young. My father told me that some hamlets in the Westhoughton area of Lancashire lost nearly all their young menfolk in the awful explosion at Hulton Colliery in 1910, when 344 died. Some families lost as many as four close relatives.

Explosions are traditionally regarded as the most serious of the hazards faced by the coal-miner in the course of his daily work. Traditionally, firedamp is blamed for colliery explosions, but in fact, with rare exceptions, violent explosions have always been caused by the combustion of coal dust. Twenty-five explosions, each resulting in the deaths of upwards of 100 persons, carried off 4,665 men and boys. Another 2,112 persons were killed in the forty-four explosions which each involved between 50 and 100 men and boys. Altogether somewhere around 12–13,000 persons met their deaths as the result of underground explosions. Some of the worst have occurred this century:

Damage to shaft fan house and other surface buildings caused by an underground explosion at Maypole Colliery, Wigan, in 1908. 75 men were killed but 3 managed to escape through old workings connected with another pit

Blundell, a Lancashire colliery owner, by William Armstrong of Newcastle-upon-Tyne, a well-known consultant viewer, speaks for itself about the destruction caused by a coal dust explosion, which the one he describes undoubtedly was. It also illustrates the immense and formidable tasks which confronted mining engineers and colliery managers after major accidents:

Testing for gas with a 'baby' Wolf flame lamp. These lamps were used only for gas testing and were not very popular

		dead
Universal Colliery,		
Senghenydd, Glamorgan	1913	439
Pretoria Pit, Hulton Colliery,		
Lancashire	1910	344
Gresford, Denbighshire	1934	265
West Stanley, Co Durham	1909	168
Podmore Hall, Staffordshire	1918	155
Wellington, Whitehaven,		
Cumberland	1910	136
William Pit,		
Whitehaven, Cumberland	1947	104

A letter written to Col Henry B. H.

13th September, 1880   Pelaw House,
                       Chester-le-Street.

My Dear Sir,

You are quite right, the rapid air currents now dealt with are more fruitful of explosion than the sluggish air of Olden time. These heavy discharges of gas at once mixed up to the inflamable point, where it is quite possible that a current of ¾ the velocity would have been insufficient to have made the mixture explosive, and it would have passed away.

If anyone can invent a lamp, which shall, after the ordinary examination and test, be extinguished, whatever the magnitude of the discharge, and yet afford a fairly good light to work with, or which shall continue burning in an explosive mixture, and yet never communicate its flame to the outside gas, he would be entitled to a place in Westminster Abbey.

There are plenty of lamps which are extinguished, but then they are complicated, in many parts, and as working pitmans lamps become unsafe. Then, so long as glass is used, there is a chance of a pick breaking through, or the glass may break through heat expansion. All these are vexed questions which obstruct the acquisition of a perfect lamp. Light is essential, and gauze absorbs it too much. But give me a lamp in two or three parts, which can't be tampered with, easily locked and opened, easily cleaned and easily examined, and which under all conditions of gas mixture is extinguished, and explosions I feel certain must be numbered on the fingers of your hand.

On Saturday night I left 135 (or there abouts) dead men in the pit at Seaham, the entire colliery a gasometer, the furnace out, and with nothing but natural ventilation.

Two hundred horses and ponies in the stables, all rotting and creating a stench utterly impossible to conceive, with no prospect of recovering a single body for some week or 10 days, and you may conceive the business there is before us.

In all my experience, and it has been an extensive one, in explosions, I never saw so great an exhibition of force and destruction. Twenty four legs have been passed without a body, and whether the legs of one man or two men we cannot say. A body lying beheaded, with the ghastly skull from it, and every corpse stark naked, every rag of clothes blown straight off them and scattered about the pit. We took up the last man within reach last night.

I have improved our 'Pelton lamp' I think considerably, you have many of them at Pemberton, and experiments are being tried every week, hoping to solve the problem.

I am, my dear sir,
Yours faithfully,
Will Armstrong.

In early childhood, the writer heard the story of the Greener, Watkin and Pickering families – all related – who managed the Pemberton Collieries in Lancashire. William Greener Snr, a Durham man, was killed by a roof fall in 1865. William James Laverick Watkin, also from Durham, met his end with two of his officials in a brave attempt at rescue after an explosion in 1877, which had caused much destruction and the deaths of 33 men. Watkin and his officials died from carbon-monoxide poisoning, as at that time no efficient breathing apparatus had been invented. Watkin's stick and leather helmet were kept in the general manager's office, and I remember experiencing a morbid interest in seeing them and also in passing the sealed-off district in the pit where the explosion took place.

Greener's son, William John, was killed in 1897 by a fast train of tubs in one of the pits. My father, who was then a young apprentice, saw his mangled body being carried to his home. His widow was left with nine children, the eldest 16 years of age. Her brother, Harry Pickering, trained at the colliery and became the Chief Inspector of

CADEBY COLLIERY, 1912

In the midst of great rejoicing, to welcome both our King and Queen
Came a blow of stinging sharpness, changing all that joyful scene.
Came the news, 'A Mine explosion', men are missing'-then the cry
Of the noble rescue party - we are here to do or die!

Came the news, a little later, shocking news and sad to tell-
The rescue party, noble heroes, had perished-how our hearts do swell
When we think of all their dear ones left to sorrow midst the joy.
Sorrow for their Father, Husband, lover or an only boy.

Sympathy throughout the Kingdom flows alike from high and low.
Royal hearts and hearts of Miners bow their heads in deepest woe.

A memorial card of the Cadeby Main Colliery explosion. 88 were killed in two explosions, the rescue team led by W. H. Pickering, Chief Inspector of Mines, being lost in the second one

Mines in charge of the Yorkshire and Lincolnshire District. He was also killed while leading a rescue party, at Cadeby Colliery after the explosion there in 1912. Such stories are told in scores of colliery villages in every coalfield.

What set the miner apart from most other industrial workers was the risk to life and limb that he took in earning his daily bread. In the nine years that the author worked at Pemberton Colliery, Lancashire, the following accidents took place which can be taken as typical of any colliery employing about 2,000 men fifty years ago: in seventeen fatal accidents, eight were crushed to death by large stones falling from the roof, six were crushed or run over and mangled by tubs on the haulage or pony roads, one had his legs cut to pieces by a coal-cutting machine and two were killed on the surface. Of these men, four were under 20, one being a boy of 15 and another 17.

In addition, there were fifty-five very serious accidents causing fractures of the skull, spine, pelvis, legs, arms and ribs, etc. Twenty of these accidents occurred to boys of 20 years of age or under. Twenty-five of the accidents were caused by falls of roof or sides, eighteen by being crushed by tubs in haulage or pony roads, four by shotfiring and six by machinery; the remaining two were surface accidents. The author knew many of these men personally and what impressed him most was the fact that their workmates and colliery officials would take any risk in order to try to save them.

There were also nine serious, reportable occurrences, including three inrushes of water and four shaft accidents: an overwind with 17 men in the cage, a winding rope slipping off a conical drum with 13 men in the cage who were trapped in the shaft for 3½ hours, a broken guide rod and a broken cage chain. There was also a serious fire underground. The grief, distress and heartbreak caused can easily be imagined from the above examples; they help to explain the close nature of mining communities.

Although Seymour Tremenheere, the

61

N... ...d QUARRIES Form No. 21.
(Revised May, 1924)

Accident, No.................. ...... 19........
(Number to be filled in by the Inspector.)

OFFICIAL EDITION.

PRESCRIBED

FORM OF NOTICE OF ACCIDENT OR DANGEROUS OCCURRENCE

(See Instructions overleaf.)

1.—Whether Mine or Quarry .. Name of Mine or Quarry .. County	~~Pemberton~~ *Mine* *Lancashire*
2.—Name and Postal Address of Owner (Occupier) ..	Pemberton Coll⁵ Co (1929) Ltd Wigan.

3.— Names of Persons killed or injured.		Age last birthday.	Ordinary Occupation.
Killed.	**Injured.**		
Died 4-45 on the 17th	Jno. Johnson	15	Pony Boy

4.—Date and hour of Accident or dangerous occurrence ..	15th March 1933 @ 2-20 pm
5.—In which hour of the Shift Accident happened, *e.g.*, 1st, 2nd, 3rd, &c. [If the injured person did not begin or end work at the beginning or end of a Shift, state the number of hours he had been at work] ..	The 8th
6.—Place of Accident or dangerous occurrence	1st South Dist pony road some 25 yds from the face. *Wigan 9 feet*
7.—Cause and Description of Accident or dangerous occurrence .. .. (in case of injury by explosive, state name of explosive and how fired.)	Not clear as to how the accident occurred. The Boy says he was hooking his pony to the Box & that The Pony started and he was caught against a Bar Riding is not permitted at this portion of the road
8.—Nature and extent of personal Injury caused ..	crushed Chest and Punctured Lung
9.—Residence of injured person	10 Vere St .. Wigan
10.—Place to which injured person has been removed..	Infirmary Wigan

Mr Coalworth Hospital (vertical marginal note)

(Signed) Thos Cook .... Owner Agent Manager.

Date 15th March 1933

To W. J. Charlton Esq ...., H.M. Inspector of Mines.
Prudential Buildings
King St Manchester.

(113ss) (329822) Wt. 21799/1650 Gp. 144 10000 1-26 W & S Ltd.

A rescue team at the Lancashire Central Rescue station at Howe Bridge in about 1920. This rescue station was established by the Lancashire and Cheshire mine-owners in 1908

first Inspector of Mines, had been appointed in 1843, it was not until the Mines Act of 1850 that the principle of state intervention in the interests of safety in the mining industry was established. By 1852 six inspectors had been appointed. Mining disasters occurred regularly, and as the industry developed in size the number of accidents increased in proportion. New mining safety laws were introduced from time to time, and more inspectors were recruited. By the end of the century about 30 had been appointed, but following the passing of the Coal Mines Act of 1911, the figure rose to 90, and different grades were established. By the 1950s there were about 165 inspectors, including those for electrical and mechanical engineering, medical examinations, pit pony welfare and for the inspection of quarries. Their duty was mainly to ensure the carrying out of the 220 sections in the Mines & Quarries Act and the 600 regulations made under them. These cover every aspect of safety, including statutory daily, weekly and other reports, mine plans, lighting, signalling, telephonic communication and other electrical and mechanical apparatus, shafts and shaft equipment, roadways, locomotives, track, conveyors and the percentage of combustible dust in the roadways. Combustible dust is treated by limestone dust and must be kept below 50 per cent.

Other serious aspects of mining legislation are blasting, which is strictly controlled by no fewer than 76 regulations; roof supports, which, especially on the faces, were mostly timber until the 1930s, the colliers at first being prejudiced against steel supports, as when a roof collapse was imminent, timber started 'ticking' or cracking under

A stone dust barrier. In the event of an explosion, the pressure wave leading the blast will dislodge the shelves and a dense cloud of stone dust is formed, which extinguishes the explosive flame

pressure and often gave time for the men to move to a safer place. Most of the early steel props gave way suddenly under a heavy weight — as the colliers said, they were 'flirted' out. The modern hydraulic supports on a coal face nowadays make conditions infinitely safer for the face worker, and steel arches which support main roads are a tremendous improvement on the old timber supports.

Ventilation has been greatly improved over the years. Even during the early part of this century, some collieries were still dependent upon furnace ventilation, whereby a large furnace was made at the bottom of the upcast shaft which altered the density of the air by heating it and thus caused it to rise up the shaft. Although the main, sometimes gas-laden, air current was generally carried into the shaft through a 'dumb' drift and not over the furnace, it was not altogether free from danger. Large fans at the surface of the upcast shaft had been introduced in the 1860s and this is the system used today. At a large, now closed, pit known to the writer, the main fan produced 600,000cu ft per minute. An adequate amount of ventilation is that which will dilute and render harmless inflammable gas, prevent the amount of carbon dioxide being increased to more than 1¼ per cent and

A roadhead in a seam. It is the main intake and the main conveyor road. Note the stilts at the foot of the steel arches; these 'close' as the roof subsides

prevent the amount of oxygen in the air being reduced below 19 per cent.

At a coal face, methane is given off from the roof, the floor and the coal itself, and in a gassy seam it is necessary to remove the gas from the roof and the floor before it can enter the air stream. To deal with the problem boreholes are drilled into the roof and sometimes into the floor at a frequency determined by the amount of gas to be dealt with. At the colliery mentioned above, some of the floor holes tapped gas at a pressure of up to 80lb/sq in. Rock bursts, accompanied by large volumes of methane, occurred in a tunnel during drivage.

The gas from the drainage holes was piped up the pit to exhauster pumps and used to fire seven Lancashire boilers.

Apart from accidents, the miner was prey to malignant diseases peculiar to his job. He could be striken with pneumoconiosis or nystagmus, beat knee, beat hand or other less virulent but crippling complaints. Pneumoconiosis is a deadly disease of the lungs caused by the constant inhalation of dust, especially silicious dust. It makes the victim prematurely old and brings about an early death. During the period 1910–12, 743 miners died from pneumoconiosis, but many times that figure were disabled by it.

Nystagmus, which was quite common until the 1930s, results in a rapid involuntary oscillation of the eyeballs; it

Above: The colliery lamp room at Gibfield Colliery, Atherton, in 1906

Below: Man riding. A deputy on his rounds along the low roads at Old Meadows Colliery, Bacup, Lancashire. He is pushing off the floor with his right foot, with his left knee on the trolley base

is an extremely painful disease, causing temporary blindness, and is peculiar to men who work underground. It is caused through working with a feeble light in a cramped position, and constantly straining the eyes by trying to see the cleats in the black coal when striking with the pick. When one considers that the original type of Davy Lamp gave about ⅛ of a candle power, it is no wonder that the colliers of old days used candles surreptitiously, despite the hazard of gas. The more modern Marsaut lamp seldom gave out more than half a candle power. Whitewashing and stone dusting with limestone dust, only brought in early this century, helped to alleviate the situation. Cases of nystagmus for which compensation was paid rose from 1,618 in 1910 to 7,028 ten years later.

Beat knee is similar to housemaid's knee and is caused by crawling about on all fours on low coal faces, and kneeling whilst getting or filling coal. Knee pads to protect the knees only came into general use in the early part of this century, but the early home-made types were sometimes less than useless, in that sharp pieces of coal often got between the knee and the pad, and made crawling about on all fours rather painful. In 1910, compensation was paid for 796 cases of beat hand and 1,128 cases of beat knee. In 1920 these numbered 762 and 1,387 respectively.

The Mining Village and the Miner

There were great differences in colliery villages. Some were in rural areas, but other mining communities were close to and integrated with industrial towns. Some colliery villages were set in bare desolate areas with very little that was beautiful to see, especially once the colliery was established with its spoil heaps, smoking chimneys and sometimes (but not always) ugly buildings. In such a colliery village, a great deal depended on the owner or company to provide the bare essentials, such as a school, a simple place of worship and perhaps a recreation ground. Workpeople employed at collieries situated on the outskirts or within the environs of a town, were better off, in that they could enjoy all the amenities a town provides. An example of such a colliery, working until recent times, was Bradford, almost in the middle of Manchester. In these cases, alternative employment to mining was available for the young members of a miner's family, as for example in the large engineering works near Bradford, Wheatsheaf and Pendlebury.

At the other extreme, the families of miners in, say, a remote valley in Wales had little choice in the matter – the colliery was often the only employer, except for odd jobs like postmen, railway station staff, etc. However, it does not seem to be recognised by the public at large, that craftsmen of almost every kind and of the highest skill are employed at collieries, and apprenticeships to the various trades were always eagerly sought after by boys who wanted something 'better' than a job in the pit.

I well remember that boys had their names put down for apprenticeships at a colliery concern, owned by a very progressive private colliery company, years before they left school. At the colliery where I served my apprenticeship as a surveyor, the trades included fitters, blacksmiths, electricians, joiners, carpenters, wagon builders, saddlers, tinsmiths, painters, sign-writers, masons, bricksetters, plumbers, etc.

In addition, jobs were available in the boiler plant, washery, brickworks, sawmills, mortar mill, loco shed, screens and as drivers of the many stationary steam engines about the colliery. A boy who was taken on to drive a small screen engine could hope eventually to be given the job of driving one of the large engines on the surface that operated underground haulages, and from that to graduate to a winding engine. Winders were always proud of their skill and their heavy responsibility, and seemed to form an élite of their own. A few apprenticeships were also available for surveyors and mining engineers, for draughtsmen and for chemists in the colliery laboratory. The latter were required for the coke and by-products works associated with the colliery, apart from analysing coal, dust and air samples from the mine. Last, but not least, came the staff of wages clerks, weigh clerks, despatch clerks, store clerks, cashiers, etc. The brightest boys from the colliery school were generally selected for training in these jobs.

In the rare cases of a colliery owner

Officials in their underground office last century, either making up the time and piecework books or the statutory shift, daily, or weekly reports

68

The overmen's underground cabin at South Hetton Colliery, Sunderland

living near the pit, a further dimension to life and opportunities for alternative work were added. The author was brought up within a mile of a large colliery employing, before the 1930 depression, 3,000 men. In the opposite direction and 200yd away was a 500 acre park, with a large Elizabethan mansion set in the middle of it and all the paraphernalia of a great country estate. We were as familiar with gamekeepers, woodmen, footmen, butlers, pantry boys, housemaids, lady's maids, linen maids and so on, as we were with winding enginemen, loco drivers, fitters, colliers, datallers, pony lads, shunt minders and pushers-on. We often saw house parties at the mansion in the park, like those every Grand National day at the end of March, when the head gardener was expected to have both strawberries and new potatoes available.

Of course estates in coalfield areas were often richer, because of the revenue from coal royalties, than many of those in the depths of a purely rural countryside, and quite often the income from coal enabled the owners to purchase secondary estates in other less-industrialised parts of the country.

There were drawbacks in having a country house and park near a colliery, apart from the smoke, dirt, noise and sheer unsightliness. For instance, more gamekeepers and assistants (both full- and part-time) were required on an estate near colliery villages to deal with poachers, of which there were many. Colliers were intrepid poachers, always ready for a fight with gamekeepers. I remember a collier, a regular night poacher, who had a gun he could take to pieces and conceal under his jacket. He told me how to catch pheasants silently in the night by various methods, including a fishing line with raisins

70

Lancashire pit-brow girls in 1925. Their job was to work on the picking belts which carried large coal at a slow speed to the wagon-loading jib. They picked out pieces of dirt and broke up, with a small pick, large lumps of coal containing dirt bands, throwing the latter out. In most coalfields, men were used for this purpose

steeped in port on the hook.

As previously mentioned, the quality of life in a colliery village depended to a great extent on the colliery owners. Sometimes, during the middle part of the nineteenth century, much of the property was built by speculators; it was often jerry-built, of mean appearance, with cramped accommodation, the front of the house on the pavement and an open unpaved yard at the back with communal 'open midden' privies. If the occupier of the house was fortunate enough to have a piped water supply, there was only one tap over a 'slop stone' in the kitchen. It must be said though, that these houses were an improvement on the former 'back to back', one up, one down houses with two rows under one roof and only one outside door to each house. A colliery manager writing in the 1920s about colliery houses at Killingworth, Northumberland, built in 1802, said:

The rooms were small and low, and the windows as a rule would not open. Sanitary appliances were non-existent. The only sewers were open, running past the cottages near the doors, so that they had to be crossed by those going in and out. No miners, however large their families, had more than two rooms, and newly married men had to be content with one small room.

'Before and after'. Scene at the doorway of a pit-head baths about 1915

Yet many of them lived to a hale old age, working regularly down the pit. In 1874 there were upwards of thirty sexagenarians working at West Moor, and several of seventy years or more were still at work. This strength and longevity were due perhaps to the fact that they had no inducement to stay indoors, and from their earliest years they spent most of their time, when not at work, in the open air. They certainly bred strong men in those days in spite of their insanitary dwellings.

Conditions for colliers, black with coal dust, to have a thorough wash in hot water, could not have been more difficult. One has only to imagine the feelings of a collier, absolutely tired out after a day's work, having to go through the laborious process of bathing. The tradition, which still persisted in Lancashire when I was a mining student, that washing the back weakened the spine, absolved the less-particular colliers from thoroughly washing the body, and I have known men who only washed their heads, necks, arms and feet during the week. Clean calico pit drawers were expected to rub off or absorb the dirt. This, of course, applied only to an ever decreasing minority fifty years ago.

An example of a colliery firm that pursued an enlightened and responsible policy towards their workers was the Atherton Colliery Co in Lancashire. The company was a private one, the

One of the earliest pit-head baths in the country, at Gibfield Colliery, Atherton, Lancashire, built in 1914

owners being members of the Fletcher and Burrows families. They built very commodious houses for their workpeople and kept them in good repair. They also provided an excellent club house, playing fields, bowling greens, tennis courts, gymnasium, etc. A joint committee of owners and men was responsible for the general welfare work at the collieries. During the early part of this century there were committees for football, bowling, cricket, male voice choir, fishing, dramatics, gardening, education and tennis. An Atherton Collieries Workpeoples Relief Fund was set up during World War I to give financial help to workmen who served in the forces, and to relieve cases of distress amongst their families.

In 1913 and 1914 they built pit-head baths at their four collieries, these being some of the very earliest in the country. A large up-to-date laundry catered for washing dirty pit clothes, and all the staff, down to colliery deputies, were provided with a good suit of clothes each year. Members of both the Fletcher and Burrows families lived near their collieries, were very able people, and took an active part in the management.

From the late nineteenth century, many good colliery houses were built in every coalfield, most of them with three bedrooms, a living kitchen, parlour and scullery. Some were built by the miners themselves, especially by those who did not spend their money on drink and gambling and otherwise lived frugally. The writer remembers a pick sharpener who built two rows of houses, the rent from which provided an income for him in his old age. Sometimes co-operative societies helped miners to build or to buy houses, especially in the north of England.

Examples of excellent villages built by colliery companies are those at Blackhall Colliery in Durham, Brodsworth Main Colliery at Rossington, Yorkshire, Maltby Colliery village at Edlington, built by the Yorkshire Main Colliery Co at Bolsover in Derbyshire, Cresswell in the same county and at Mansfield in Nottinghamshire. In South Wales a garden village was built in the 1920s near the Powell Duffryn Coal Company's Brittania Colliery in the Rhymney Valley. It remained true, however, that up to the 1920s the worst housing for miners was in South Wales and Scotland.

The NCB took over these colliery houses when the mines were nationalised in 1947, and many house improvement schemes have been carried out. Since July 1976, it has been the policy of the Board to sell their houses to tenants, and up to March 1980, 28,000 had been sold with another 4,000 in course of completion. The Board's housing stock was still 55,000 at that date.

As regards the miners themselves, their daily work in harsh repellent conditions, in darkness and dirt, in deep mines, exhausting heat, with the ever present danger of falling roof stones and poisonous and inflammable gas, sets them apart from workers in more congenial industries. Half a century ago a colliery manager wrote:

The typical miner is not always an attractive individual on the outside, but an uncouth and unprepossessing exterior often hides a strong and resolute character and a kindly disposition. Accustomed to face stern and disagreeable realities in his daily work, he is real and genuine in his feelings and conduct; he has a robust individuality of his own, is self-reliant and independent and no respecter of persons, but his respect once gained is sincere and lasting.

Some miners had an air of great dignity and went about their work with

A fine village church built in 1887 by John
Pearson and Thomas Knowles, owners of the
Ince Moss Collieries near Wigan, who contri-
buted £5,000 each. It was designed by Paley &
Austin. The colliery is now abandoned, the
church demolished and nearly all the colliers'
houses have been demolished under slum
clearance orders

Sunday School walking day in a colliery village in 1930. Clergy, like the vicar shown in the photograph, were almost always a 'tower of strength' to the workpeople during depressions with short-time working, unemployment and strikes

quiet confidence. A man of this calibre worked for the writer until he was in his eightieth year. He was capable of tackling any dangerous job without a sign of panic, and indeed his confidence inspired those working under him. He never drank during the week, but at weekends he could see the best of them under the table.

Of course there is as much variety in the mining community as in any other group of people. As a general rule, the longer established the settlement, the better its sense of community and personal relations. At newer collieries, however, where there was a desperate need for labour, some rough characters

were attracted. This must have been particularly evident during the late nineteenth century: in the forty years from 1873 to 1913, the national output increased from 128 million tons to more than twice that figure – 287 millions. It does not need much imagination to see that a tremendous number of raw recruits must have gone down the new pits in those forty years. I remember many descendants of people from all over England, Scotland, Wales and Ireland in the colliery communities near Wigan: farm labourers from counties such as Oxfordshire, Lincolnshire and Gloucestershire came as horsemen; tin-miners from Cornwall arrived as the mines in their own county closed; Staffordshire, Welsh and Scottish miners came trying to better themselves as did large numbers of Irish. People from rural counties nearly always became regular attenders at the colliery

Sunday School sports at a colliery village, held in Winstanley Park, Lancashire, which belonged to the royalty owner. The coal preparation manager is holding the string on the right of the picture and one of the under-managers is seen on the middle left

church, which was probably the only thread of continuity between them and their native village; everything else was alien to them. The Welsh, of course, soon established their chapels, and new Roman Catholic churches were built to accommodate the Irish. All these churches and chapels, with their attendant Sunday Schools, had a great civilising effect on mining communities. Walking Days, Field Treats and Sermon Sundays were important events in summer time in many northern villages. Colliery workers became Sunday School teachers, lay readers, local preachers and church committee members.

Evening classes, especially those at the local mining and technical colleges, were very popular earlier this century amongst the more earnest and ambitious young people. The courses on mining, electrical engineering, mechanical engineering and commerce were well supported, and many evening students had to attend three and sometimes four nights a week to have the benefit of the full course. Day release was seldom granted fifty years ago, except to surveying and chemistry apprentices and the few selected for mine management under the wing of the colliery agent or manager. Many others attempted the statutory certificates to qualify as under-managers and managers by taking evening courses only and working in the pit during the day. Needless to say they were often tired out when they arrived at evening school, and some of them attempted the examinations many times before they passed or gave up altogether. The writer remembers a deputy with a

The opening of a magnificent sports pavilion at a Lancashire colliery village in 1922, given by the colliery owner and built by the colliery craftsmen under the supervision of the writer's father

large, young family who took a correspondence course for an undermanager's certificate and stayed up half the night studying several nights a week. He had to be at the pit before 6.30 every morning. After failing the examination eight times, and after many years, he finally gave up. There were many like him.

In my youth, adversity, poverty, lack of good opportunities in mining and bad conditions generally in the pits drove many young men to better themselves by studying. The colliery village I knew

so well in those days produced many first-rate men. The Sunday School Superintendent's son became Dean of York, the colliery assistant surveyor became the Chief Surveyor to the National Coal Board, others were founders of Gullicks, the well-known mining supports firm, colliery general managers in different parts of the country, both before and after nationalisation, while others became directors of large engineering works, headmasters of grammar schools and principals of technical colleges.

Colliers in their leisure time in the past indulged in many sports, including wrestling, whippet racing, pigeon flying (racing), rabbit coursing, knurr and spell (a knurr was a wooden ball hit by the spell, a club-ended stick), pitch and toss, etc. Many families bought pianos and paid for their children's tuition. It is a different scene today with miners amongst the best-paid workers in the land, owning good cars, purchasing desirable properties, taking holidays abroad and even cruises, and enjoying a much higher standard of living and education than they used to.

Miners' Unions

When the pits were under private ownership, there was much truth in a colliery manager's assertion in the 1920s that, 'of all hand workers the coal-miners are the most exorbitant and unceasing in their demands, but they are the doughtiest fighters in the industrial field'. It remains true today, that although the National Union of Mine-workers is not one of the largest, it is still one of the most powerful, in that if it chooses it can bring the country to a standstill within a short time. It is now a closed shop and a very closely knit organisation, with the great traditions of bitter struggles and the resultant successes or failures of its constituent branches and their predecessors going back at least 150 years. During the many great strikes and lock-outs of the past, the mining communities learnt to suffer and survive together, by mutual support and assistance, with little or no help from anyone else, except local clergy, tradesmen, shopkeepers, farmers and sometimes even the colliery owners and royalty owners against whom they were striking.

Some excellent books have been written about miners' unions, notably those by Page-Arnott, and it is not possible here (although the subject is an extremely important one) to give more than a few brief notes on it.

Some local banding together of miners occurred in Durham and Northumberland in the eighteenth century, mainly to try to redress local grievances where the individual or individuals were powerless to do so. As early as 1792, the colliers of Wigan struck to demand what was described as 'an extravagant advance of wages. They have given them [the agents] only till tomorrow at 3 o'clock to consider of it and if their demand is not complied with, they threaten to destroy the works by pulling up the engines, throwing down the wheels and filling up the pits.' The colliery owners requested that a body of troops should be sent from Manchester in aid of the magistrates, as they were to do on many subsequent occasions when violence was threatened, and this seems to have had the desired effect.

The Combination Laws, passed in 1799 and 1800 to suppress unions, were only repealed in 1824, but during that time the colliers used the many friendly societies as a cloak for union activities. After the repeal, colliers' unions were formed in many areas, one being formed in Wigan in 1830. By 1840 county unions were active in Northumberland, Durham and Yorkshire as well as Lancashire, but there were also unions in parts of Scotland and Staffordshire.

An important step forward took place in 1842, when the Miners' Association of Great Britain and Ireland was formed. This was a federation of the county and other district unions. Delegates of the Association met in Newcastle-upon-Tyne in 1843. David Swallow of Wakefield was the first secretary, but Martin Jude of Northumberland was the driving force behind it. It rapidly gained in strength, and by 1844 there were 100,000 members from nearly every coalfield in the country.

William Prowting Roberts, a very able lawyer, was paid £1,000 a year (an enormous sum in those days), to fight its cases. Unfortunately, it dissipated its resources in supporting the Durham and Northumberland miners, who were on strike for 18 weeks in 1844, and were again in dispute with the colliery owners during the late 1840s and early '50s. From 1847, there was a severe depression in the coal trade; coal prices were much reduced, with a consequent reduction in wages. The Association, and the unions forming it, appeared powerless to the men; there was a catastrophic decline in membership and the Miners' Association of Great Britain and Ireland petered out in 1848.

District Lodges, however, carried on, but after further wage cuts were imposed, there was much agitation by the mid-1850s for an increase in wages. The resistance of the owners to the colliers' demands, resulted in rioting and violence, especially in Lancashire. Alexander MacDonald, a Scottish miner who had struggled to educate himself until he achieved his ambition of becoming a student at Glasgow University, came to England in the late 1850s and eventually founded the Miners' National Association. A rival federation, the Amalgamated Association of Miners, was founded by the miners of Lancashire and Cheshire in 1869, and soon attracted a large membership from South Wales, where the miners' wages were considerably lower than in Lancashire. Other associations were revived at this time, but eventually they polarised into the two federations. In the meantime, when the Employers' Liability Act came into force on 1 January 1888, many of the owners gave notice that existing contracts of service would cease at the end of 1888, and that the new contracts would require them to contract out of that Act. This they would not do, and after a week or two the owners gave in.

By 1908, all the district unions were affiliated with the Miners' Federation of Great Britain, which had been established in 1889, and this became the negotiating body with the mine-owners and the Government in respect of wages, hours of work, recruitment, etc. Before its formation, some of the county unions, including the Lancashire Miners' Federation, had agreed to a sliding scale whereby wages and field prices rose or fell 2½ per cent for every 4d variation in selling price.

It soon became apparent that those districts not affiliated to the MFGB, but still under the wing of the old Miners' National Union, were much worse off as far as wages were concerned, than the former, especially in the important coalfields of South Wales, Durham and Northumberland. This made a big difference in the cost of production between the federated districts and the rest, and the coal-owners, taking advantage of a trade recession in 1893, demanded a reduction of 25 per cent in the wages of miners in the former. The Federation resisted, and this resulted in 300,000 of their men being locked out from 28 July to 7 November. They were supported, for much shorter periods, by men from parts of South Wales, Cumberland, Fife, Kinross and west Scotland. The Cumberland men were only on strike for five days.

The writer's grandfather, who at that time employed between thirty and forty men on tunnelling, ripping and sometimes pit-sinking work, had to disband his men, and since he had no other income he went to work at the William Pit at Whitehaven. His children were not allowed to stay in school at lunch times and partake of the potato pie or hot pot provided by the strike committee, which was only natural. Although there was a great deal of suffering and distress, soup kitchens, serving soup, hot pot and potato pie,

Preparing food for strikers' families in the 1921 strike

were set up in every district, in Sunday Schools, chapels, public houses and so on; old people were fed at home and food tickets were given to needy people so that they could obtain supplies from shopkeepers. Milk, meat and potatoes were given by many farmers and shopkeepers who allowed the miners' wives considerable credit. Even some of the colliery companies gave food tickets and ran soup kitchens, and royalty owners were known to feed miners' families and give money to the strike fund. There had only been two weeks' strike pay for the men, and people were asking where all the money paid in by the miners had gone. The Bishop of Liverpool, Dr Ryle, stirred up resentment by describing some miners leaders as 'fluent agitators who had nothing to lose by mistakes or strikes'. The men won their battle and went back on the old rate of wages. Instead of the sliding scale, a Conciliation Board of an equal number of coal-owners and miners' representatives was set up to determine from time to time the rate of wages.

The above description can be applied in a general way to the conditions which prevailed in the miners' strikes which followed in 1912, 1921 & 1926. The Coal Mines Regulation (8 hours) Act had become law in 1908. Previously very long hours had been worked in some districts. At many pits in Lancashire, the shift was from 10–12 hours a day at the turn of the century. Men working in abnormal places – where there were bad roofs, faults, steep inclines or wet workings – had to depend on the undermanager to make their wages up to a reasonable amount. This he often refused to do, on the assumption that if he did so they would not try to produce as much coal as they could. This led to agitation for a minimum wage. The

Mounted police on duty during the 1912 strike for a minimum wage. The 'military' were made much use of during this strike, the biggest ever seen up to that time, when over a million men were out. Note the typical colliers' houses of early this century in background

Government introduced a plan which was wrecked by the coal-owners of South Wales and Scotland, and on 26 February 1912 there began the biggest strike Britain had ever witnessed, a million men being out. On 29 March, the Government passed the Coal Mines (Minimum Wage) Act. This guaranteed 6/6d a day for a collier. The men went back on 15 April. There had been violence in some coalfields and the police had called for military assistance.

Output of coal from British pits reached its peak in 1913 at 287 million tons, when coal was Britain's largest industry, apart from agriculture, with over 1,100,000 employed. By the end of the war, output had dropped to 230 million tons, and exports had dropped from 77 million tons to 28 million tons during the same period.

In 1919 the MFGB, through its President, Bob Smillie, made the following demands:

1 30 per cent increase in wages
2 A six hour day by legislation
3 Nationalisation of the mines
4 Full standard War Wages for any demobilised miners who may not be immediately employed, and full wages to any men who entered the mines during the war should they be displaced to make room for demobilised soldiers

Lloyd George then appointed a commission under Viscount Sankey which recommended nationalisation of the mines. Output per manshift dropped and absenteeism increased, due partly to men trying to avoid paying income tax and partly due to returned soldiers not being equal to working eleven days in a

83

COAL at 4/5½ per ton

PITY THE POOR MINER !

The Miners' wages have risen by 157 per cent. since 1914.

Now the Miners want another 2s. a shift, and Mr. Smillie is to plunge us into a National Coal Strike unless they get it.

We all want the Miners to be well paid and comfortable; but in return we look to them for coal at fair and just prices. If *they* get too much in wages *we* shall have to pay too much for COAL. And the vast majority of the Nation's workers have much less money than the Miners get.

Further, the cost of coal is not very serious for the Miners.

They can get as much as they want at a cost that is merely nominal. Very large numbers get their coal absolutely free. Others have to pay a trifling sum for it. The Miners consume 6½ million tons a year at an average cost of only 4s. 5½d. per ton.

This new Coal Demand must be resisted, because we who are not Miners cannot afford it, and the Miners are so well paid already that they have no right at all to claim it.

Unhappily there is another grave reason against it.

EVERY NEW WAGE CONCESSION MEANS LESS COAL.

The more money the Miners get the less Coal they give. The truth is they want and take more time to spend their wages.

P.T.O.

Propaganda against increases of wages for miners ,
after World War I

OFFICIAL FIGURES PROVE IT.

Following on the first Wage Increase (September 17th, 1917), the average output fell 6.3 per cent.

Following on the second Wage Increase (June 28th, 1918), output fell a further 6.6 per cent.

Following on the third Wage Increase (March 20th, 1919), output fell a further 3.1 per cent.

Following on the fourth Wage Increase (March 12th, 1920), output fell a further 6.6 per cent.

In all, after four Wage Increases, the ratio of output per unit of labour has fallen 22.6 per cent.

Do we need Coal? Are we short of it?

If we really do need it, are we, in face of these unchallengeable facts, to give the Miners still more money so that they can give us

STILL LESS COAL?

PRINTED BY A. & E. WALTER, LTD. 13 TO 17 TABERNACLE STREET. E.C.2.

fortnight. On 16 October 1920 they struck for an advance in wages and their demand was met. But in 1921 the Government, which had controlled the mines since the middle of the war, handed them back to the owners. They, faced with declining markets at home and abroad, immediately instituted drastic wage cuts. The result was a strike from 1 April to 1 July, which ended in the defeat of the miners and a reversion to separate wage bargaining in each district.

The writer was a very small boy during the hot summer that the strike occurred. Near my home there was a wood in which a seam outcropped. Before long, many holes had been dug by the strikers into the crop of the seam and there were even shafts about 4ft square up to 20ft deep. I was given a job in one of these shafts, hooking a bucket on to the rope at the bottom. The local branch secretary worked one of these small, illegal pits. Small trees were felled and branches cut off to provide pit props, but some of the greedier men found a lucrative market for the large beech trees, in a Liverpool mangle roller manufacturer. The landowner attempted to control all this by issuing permits to the strikers, but they were in no mood to knuckle under to anyone. One heard occasionally of 'the mob', and one day it appeared near the wood where the other strikers were getting coal, with the intention of turning them out. Those in the wood had heard the mob was on its way and were prepared for them, lined up like a regiment with picks in their hands. The mob was forced to retire in disarray and never reappeared. They did much damage, however, by throwing tubs off pit brows, smashing engine house windows, etc.

A temporary boom, which followed the French occupation of the Ruhr, soon came to an end, and in 1925 the coal-owners demanded an increase in hours from seven to eight and a reduction in wages. The Government, alarmed at the situation, provided a subsidy and a Royal Commission was appointed to review the industry, under Sir Herbert Samuel. It recommended against nationalisation and against the continuance of the subsidy from the tax-payer which was equivalent to 2s 6d per ton. It recommended a reduction in wages and/or a lengthening of hours. The Mineworkers' Federation refused to accept these proposals and on 1 May 1926, the day after the subsidy ended, the stoppage began. For nine days the miners were joined by every other union, in the General Strike. The negotiation of district agreements (the owners refused negotiation on a national basis) started on 23 November, and all pits were at work by 22 December. The miners having surrendered, the wage cuts were implemented which resulted in wages falling by more than a shilling a shift, and hours were increased to the pre-war standard.

I well remember on my way home from school in October 1926, witnessing a baton charge by the police (most of whom were not members of the local force) who dispersed a hostile mob of 2,000. At the local colliery 300 miners had commenced work, and the crowd had gathered to accost and it was feared offer violence to them on their way home. The crowd refused to disperse, whereupon the police charged them with their batons; several men received injuries to the head, but none were taken to hospital. Some of the men who went into a field threw stones at the police.

Conditions of employment at a large colliery in the 1930s

Overleaf: Strikers and their families picking coal from the tip at a Lancashire colliery in 1893, when the stoppage lasted for 16 weeks

PEMBERTON COLLIERY.

CONDITIONS OF SERVICE

BETWEEN THE
Pemberton Colliery Limited,
AND THE
Workpersons Employed by them
AT

Pemberton Colliery, Wigan

1.—Unless prevented by sickness or accident, every collier shall (unless otherwise required) work eleven fair days' work in each fortnight; and every fireman and other workman twelve days (unless otherwise required).

2.—Every person employed shall give or receive fourteen days' notice, in writing, before leaving the employment of the said firm. Any person guilty of theft, fraud, or gross misconduct is liable to instant dismissal by the Manager. All notices shall be given on the making-up day.

3.—In case of an accident at the Collieries preventing any collier or other workperson from attending to his work the contract of service with any such collier or workperson shall terminate, and the condition No. 2 shall not apply.

4.—Every workperson shall be responsible for the materials and tools entrusted to him, and should he lose, destroy, or allow to be lost or destroyed any of such tools or materials, he shall make good the same, or pay the value thereof.

5.—Every collier, on commencing work, will be supplied with a copy of the Abstract of the Coal Mines Act, 1911, and of the General Regulations in force at the Colliery, a set of tallies, and one or two iron scotches, for which the sum of one shilling will be held off his wages, which, on leaving the firm and returning intact the above-mentioned copy of the Abstract and Special Rules, set of tallies, and scotch or scotches, will be paid to him.

6.—Every underground workperson will be provided with a safety lamp properly cleaned and trimmed with oil and wicking, for the use and provision of which he shall be liable to a deduction of the sum of one penny per shift from his wages.

By Order of

PEMBERTON COLLIERY LIMITED.

E. Sidebotham, Printer, Wigan.

ATHERTON COLLIERIES.

TO OUR WORKPEOPLE:

WE HAVE MADE YOU AN OFFER TO CONTINUE THE SEVEN AND A HALF HOUR DAY AND PRESENT WAGES TILL SEPTEMBER 1927.

WE MAKE THIS OFFER TO HELP THE MINERS IN LANCASHIRE TO A SETTLEMENT BY GIVING A LEAD.

THIS IS A DEFINITE POLICY!

HAS THE FEDERATION A BETTER ONE?

ASK THE SPEAKERS
AT THEIR NEXT MEETINGS

WHAT IS THEIR POLICY?

FLETCHER, BURROWS & CO. Ltd.

October 19th, 1926.

HUNT & MANSLEY, Printers, Atherton. Tel. 147.

In 1927, only the South Derbyshire, Leicestershire, Cannock Chase, Warwickshire and Yorkshire coalfields made profits, the latter only being ¾d per ton. During the following eight or nine years, until rearmament, the coal industry was in the doldrums, with short-time working at most pits until the mid-thirties. The coal industry also had its share of the tragic unemployment of those days: 355,000 miners being out of work in 1932, many more than work in the industry today. The trouble was excess capacity and the resultant cut-throat competition of hundreds of separate companies.

The Coal Mines Act of 1930 was fairly successful in its intention of regulating prices and production, but the other parts of the Act – setting up a National Wages Board, and a Reorganising Commission to promote amalgamations of colliery companies – were failures due to the non-cooperation and unwillingness of the coal-owners. Rearmament during the late 1930s created a larger market for coal, and during World War II the Government again took control of the pits. Young men were given the option of going into the forces or down the pit; those who chose the pit were known as 'Bevin Boys'.

In 1945 the National Union of Mineworkers was formed from the MFGB and all the constituent unions. From its formation, the NUM was the only body with the power to negotiate wage settlements and conditions of work, whereas after the 1926 strike, the owners ignored the Miners' Federation and would only deal with their district unions. The NUM now covers industrial workers, clerical workers, weekly paid industrial staff and coke workers. Colliery deputies and managerial staff have their own unions.

After nationalisation, there was a great demand for coal, and 224 million tons were produced in 1957. However, within a few years the situation was completely changed; cheap Arab oil was taking over from coal and uneconomic pits were being closed by the score. Power loading was introduced on the faces, and as it took over, the National Power Loading Agreement on wages accompanied it.

By 1972, miners' wages were falling behind others in the industrial league table and they put in a claim for a pay rise which was refused by the Government, who at that time had a statutory incomes policy. A committee of enquiry under Lord Wilberforce was appointed to deal with the problem, and its findings were accepted by the miners.

In 1973–4 there was another conflict when the NUM asked for a further pay award, without results. Overtime was banned and the Government, alarmed at the situation, put all industries on a three day week. Unsatisfactory negotiations caused the miners to ballot, and they voted for a national strike. Edward Heath, the Prime Minister, then called a General Election to obtain a mandate from the people to bolster his resistance to the miners, but he lost it. Mr Wilson, the new Prime Minister, immediately authorised the NCB to negotiate with the miners, in fact to meet their demands.

As regards the future of the coal industry, it is impossible to improve on a letter from Sir Derek Ezra, Chairman of the National Coal Board, to the *Daily Telegraph* on 3 August 1981. He made the following points:

1 Productivity is rising. The increase is now exceeding four per cent a year. For an industry of our size this is a considerable achievement.

2 Our investment is paying off. We are getting up to twice our annual productivity from investment that has already matured and this could rise to five times from investment in hand.

3 Markets are again expanding. In spite of the recession we have managed to restore coal as the principal source of primary energy in Britain, and we have doubled our exports.

4 Our technology is highly advanced. In many respects the progress made in Britain in the ways in which coal is produced and used have established a lead over what is being done elsewhere.

5 Our overseas earnings are growing. This results from the expansion of our exports of coal and coke and from the sale of technology and equipment undertaken by the Coal Board, their associates and suppliers. Something like £500 million is likely to be earned from these operations in the current year, making us one of the biggest overseas earners in British Industry. We are highly regarded abroad.

Glossary

Bucket pump A pump containing a piston fitted with 'flap' valves working in a vertical barrel

Bunton or **baulk** A wooden beam in a shaft

Canch or **ripping** Stone taken from the roof or floor of a seam in roadways in order to make height

Capstan engine Engine geared to lift heavy weights

Chocks Pieces of hardwood approx 6in sq × 1ft 6in long, built cross-wise 2×2 to form supports for the roof

Clack A non-return valve used in the suction and delivery of pumps

Cleat A vertical facing or parting in a seam in two directions generally at right angles to each other

Dataller Skilled miner paid by the day who carried out roadway repairs, and other general work

Downcast shaft The shaft down which fresh air is drawn

Drift A tunnel or a road in or to a seam

Fault A geological dislocation in the strata

Flat rods Horizontal rods generally at the surface which transmit the motion of a pumping engine

Gin A winding or other machine generally worked by horses

Goaf, gob or **waste** The collapsed excavated space behind the coal face supports

Inbye Towards the coal face

Jib (of coal-cutting machine) The flat steel arm around which travels the cutting chain with its picks

L-legs Quadrants at the top of a shaft where the horizontal motion of the engine flat rods is transmitted to the vertical pump rods in the shaft

Main and tail Haulage engine with two drums where the ropes connect round pulleys to the front and rear of a train of tubs. Used on undulating or flat roads

Mouthing Entrance to a seam or tunnel from a shaft

Outbye Towards the pit shaft

Outcrop or **crop** The position where a coal seam reaches the surface

Parting A horizontal break in a coal seam often filled with dirt or other foreign matter

Plug rod A vertical rod attached to the beam of an engine which operates through tappets the steam and exhaust valves of the engine

Pump or **spear rods** The vertical rods in a shaft, attached to the end of the engine beam, which actuate the pumps

Shaft pillar Coal left for the support of the shaft

Straitwork Narrow stalls in pillar work

Sough A drainage tunnel

Tappets Levers operating the valve gear of a steam engine

Upcast shaft Shaft through which foul air is expelled from the mine by the use of a fan or (in former times) a furnace

Chronology

1698 Invention by Thomas Savery of a non-reciprocating steam pump

1710 Bensham (Newcastle) explosion – about 80 killed

1712 Thomas Newcomen made his first steam beam pumping engine at Dudley Castle, Staffordshire – the first practical steam engine

1758 Duke of Bridgewater opened his canal at Worsley, Manchester

1760 A Newcomen pumping engine provided water to drive a reversible water-wheel winding gear for a mine

1763 Newcomen engine first used for winding coal at Hartley Colliery, Northumberland

1765 Strike of miners in Northumberland

1767 Smeaton improved the construction of pumps, in particular the pump bucket

1769 Watt made his first experimental steam pumping engine at Kinneil

1780 James Pickard patented the crank for converting reciprocating into rotary motion in steam engines

1792 Strike by Wigan miners

1799–1800 1st and 2nd Combination Acts, suppressing, amongst other things, associations of workmen/unions

1800 Buddle improved furnace ventilation

1800 National output of coal – approximately 10 million tons

1806 Cornish boiler introduced by Trevithick

1807 First mechanical ventilator installed by John Buddle

1810 Durham and Northumberland miners' strike

1812 First successful steam locomotive, at Middleton Colliery

1813 Sunderland Society formed, resulting in Davy, Stephenson and Clanny's invention of the safety lamp

1815 Heaton Colliery, Northumberland, innundation of water – 90 killed

1816–18 Acute distress. Violence and strikes in Lancashire

1824 Combination Acts repealed

1824 Ayrshire miners' strike

1825 Stockton & Darlington Railway opened. Northumberland & Durham Colliers' Union formed

1831 William Bickford introduced first safety fuse for mining

1831–2 Strikes by Northumberland and Durham miners

1834 New Poor Law

1835 Wallsend Colliery explosion, Northumberland – 102 killed

1835 Parliamentary inquiry into mine disasters

1839 Formation of the South Shields Committee to investigate causes of accidents in mines

1841 National Charter Association. Miners' Association of Great Britain and Ireland formed

1842 Commission on labour of women and children in coal-mines; resulting in women, and children under 10 being banned from the pit

1843 First Government Inspector of Mines appointed

1844 Strike by Northumberland and Durham miners

1844 Fairbairn introduced the Lancashire boiler

1844 Haswell Colliery explosion, Northumberland – 95 killed

1845 Compressed air introduced underground

1849–53 Strikes by Northumberland and Durham miners

1850 First Government inspector appointed with power to go underground. An act was passed in 1855 outlining powers and duties of Inspectors

1850 National output of coal approximately 70 million tons

1853–4 Ince Hall Colliery explosions, Wigan – 147 killed

1856 Cymmer Colliery explosion, Rhondda – 114 killed

1857 Lundhill Colliery explosion, Yorkshire – 189 killed

1859 Worthington invented his Duplex steam pump

1860 Black Vein Colliery explosion, Monmouthshire – 142 killed

1860 A further Act regulating inspection of mines

1860s Guibal ventilating fan introduced into Britain

1862 Hartley Pit disaster, Northumberland – 204 men suffocated. There was only one shaft. Act of Parliament passed making it compulsory for every colliery to have at least two shafts

1866 Oaks Colliery explosion, Yorkshire – 361 killed

1867 Ferndale Colliery explosion, Pontypridd – 178 killed

1872 Coal Mines Regulation Act (a comprehensive Act to regulate methods of working)

1875 Swaithe Main Colliery explosion, Yorkshire – 143 killed

1877 Blantyre Colliery explosion, Lanarkshire – 207 killed

1878 Wood Pit explosion, Haydock, Lancashire – 189 killed

1878 Abercarn Colliery explosion, Monmouthshire – 268 killed

1880 Employers' Liability Act

1880 Black Vein Colliery explosion, Monmouthshire – 120 killed

1880 Seaham Colliery explosion, Durham – 164 killed

1882 Babcock & Wilcox water-tube boiler installed at Hamilton Palace Colliery, Lanarkshire

1882 First application of electricity underground was at Trafalgar Colliery, Forest of Dean, to drive a pump

1883 An electrically driven endless rope haulage was installed at Nostell Colliery, Yorkshire

1885 Clifton Hall Colliery explosion, Lancashire – 178 killed

1887 Coal Mines Regulation Act (this Act widened the scope of the 1872 Act)

1889 Miners' Federation of Great Britain formed

1890 Llanerch Colliery explosion, Monmouthshire – 176 killed

1892 Norths Park Slip Colliery explosion, Glamorgan – 112 killed

1893 Combs Colliery explosion, Yorkshire – 139 killed

1893 Strike of Federated Districts against reduction of 25 per cent in wages, 300,000 men out for 16 weeks

1894 Albion Colliery explosion, Glamorgan – 290 killed

1894 Coal Mines (Check Weighers) Act

1895–1900 Successful electric- and compressed air-driven coal-cutting machines introduced

1900 First large electrically driven centrifugal mine pumping plant installed

1900 National Output of coal, 250 million tons

1901 Parson's AC turbo-generator installed at Ackton Hall Colliery, Yorkshire

1905 First belt conveyor introduced

underground at Glass Houghton Colliery, Yorkshire

1905 National Colliery explosion, Glamorgan – 119 killed

1905 Coal Mines (Weighing of Minerals) Act

1907 British Navy started the change from coal to oil

1908 Coal Mines Regulation (8 hours) Act (an Act limiting the miner's shift to 8 hours)

1909 West Stanley Colliery explosion, Durham – 168 killed

1909 Miners join the Labour Party

1910 Britannia Colliery, Monmouthshire – first all electric colliery in Britain

1910 Wellington Colliery explosion, Whitehaven – 136 killed

1910 Hulton Colliery explosion, Lancashire – 344 killed

1911 Coal Mines Act (a comprehensive Act replacing most previous Acts but sections of the 1887 Act were unrepealed)

1912 Miners' strike for a minimum wage – 1 million men out from 26 February to 15 April

1912 Minimum Wage Act

1913 Universal Colliery explosion Sengenhydd, Glamorgan – 439 killed

1913 Highest output of coal ever achieved in Britain – 287,430,473 tons

1918 Podmore Hall Colliery explosion, Staffordshire – 155 killed

1921 Coal mines dispute. Strike lasted 3 months – 1,100,000 men on strike

1923 & 1925 Workmen's Compensation Acts

1926 Mining Industry Act (an Act facilitating the working of mines, establishing a Miners' Welfare Fund, regulating recruitment and providing for a levy)

1926 Coal Mines dispute and General Strike – 1,075,000 miners on strike from 1 May to October/December

1930 Coal Mines Act (to regulate prices and production and to promote amalgamation)

1931–4 Mining Industry (Welfare Fund) Act

1934 Gresford Colliery explosion, Denbighshire – 265 killed

1938 Coal Act (an Act for the unification and nationalisation of mining royalties)

1943 Workmen's Compensation Act

1944 The Coal Mining (Training & Medical Examination) Order

1945 National Union of Mineworkers formed

1946 Coal Industry Nationalisation Act (an Act to establish public ownership of the coal-mining industry and ancillary activities. Provision in the Act for small private mines)

1947 1 January. Nationalisation of Coal Mines

1947 William Pit explosion, Whitehaven – 104 killed

1950 National output of coal 204 million tons

1954 The Mines & Quarries Act (an Act bringing up to date and consolidating previous Acts)

1966 Aberfan – 144 people were killed, including 116 children, when the colliery dirt tip slid down the steep hill side on which it was situated, engulfing several houses and the Infant and Junior schools in the valley below

1969 Mines & Quarries (Tips) Act

1974 Strike for pay increase after negotiations with Government broke down

1980 National output of coal – 123.3 million tons